AMERICAN SCOUTING SERIES

SCOUTING IN THE WILDERNESS
THE FORT IN THE FOREST

BY

EVERETT T. TOMLINSON

AUTHOR OF "SCOUTING ON THE OLD FRONTIER," "STORIES
OF THE AMERICAN REVOLUTION," "SCOUTING WITH MAD
ANTHONY," "SCOUTING ON THE BORDER," ETC.

D. APPLETON AND COMPANY
NEW YORK :: 1924 :: LONDON

"THE MAN HAD LAID ASIDE HIS KNITTING AS HE
APPROACHED ONE OF THE LOG-HOUSES."

PREFACE

WHENEVER and wherever courage, self-control, determination and keenly developed mental faculties are manifest, naturally they command the interest and admiration of others. There are no more enthusiastic admirers of these qualities than American boys. In the days when the contests were with wild beasts and savage Indians, who seemed to be unwilling to be driven from their home lands, and with white enemies, who were only a little less savage than the red men, these qualities were demanded of every sturdy settler. While the men were felling the forests and subduing the wilderness they were compelled to meet not only opposition but also the bitter enmity of white men and red, who were determined to gain and retain control of the region.

There is always a special romance connected with the beginnings of things. Nowhere else is this more true than in the early days of our country. Whatever of romance or greatness may come afterward, they never displace the charm of the earlier days. This story, originally published under the title *The Fort in the Forest,* has been reissued and included in the American Scouting

Series because of the interest in their pioneer ancestors manifested by my young readers. It is a source of pride and a tradition to be maintained, that America to-day owes much of its prosperity and freedom to the sturdy, sterling stock of the early pioneers.

The problems have changed, conditions are radically different, but there is the same demand for courage, honesty, self-help and determination that were so manifest along the border 150 years or more ago. If the interest in the romance and history of the beautiful region between Albany and Montreal, which my young readers have frequently written me they have, shall lead them to more detailed historical reading and research, and thereby they may come to place a new value upon the land we all love, my purpose in writing this story will have been achieved.

EVERETT T. TOMLINSON

ELIZABETH, N. J.

Contents

8   Contents

Illustrations

SCOUTING IN THE WILDERNESS

CHAPTER I

AN OLDEN-TIME PARTY

THE March sun for three days had been unusually warm, and the heavy snows had settled until, in spots, the bare earth appeared. The trunks of the trees, discolored and weather-beaten by the storms of the long winter, stood out against the background of the long stretches of snow that in places was banked in great drifts or extended in level fields far as the eye of the occasional traveller could see, like grim watchers of the dreary scene. Even the wind as it whistled among the leafless branches only intensified the barren aspect of the landscape, and the low beams of the setting sun failed to give even the appearance of warmth.

Along the road leading through the forest to a little settlement not far from Albany, a man in a rude sleigh, or "jumper," was driving an old white horse that was blind in one eye and had traversed the rough roadway so often that storms and sunshine, winter and summer, were evidently much the same to him. The slow "jog-trot" was steadily maintained ; the occasional lurching of the heavy sleigh with its wooden runners, though frequently driving it forward upon the horse's heels, apparently had no effect, and the same monotonous, lifeless gait was maintained up hill and down, through the forest, or in the more open regions where a low log-house was occasionally to be seen.

The man in the sleigh was old and weatherbeaten and rough as his immediate surroundings. A blue coat with yellow buttons, a scarlet waistcoat, leathern small-clothes, blue yarn stockings, a red wig, and a cocked hat made up his apparel; and on the seat beside him was his " work," which, if it was not the reality itself, bore a strong resemblance to coarse wool stockings, partly knitted, and giving evidence of having provided an occupation during some of the hours of his long and tedious journeying.

The man had laid aside his knitting as he approached one of the log-houses to which reference has already been made, and was leaning forward with keen interest listening to the sounds that issued from the building before him. Shouts and laughter rang out, and then there was a brief silence, followed in turn by a song, the words of which could be distinctly heard : —

" Green grow the rushes, O,
    Kiss her quick and let her go,
    But don't you muss her ruffle, O."

The verse ended amid renewed shouts and a
sound as of people rushing about. A smile of interest and of sympathy appeared on the face of the
lonely traveller, growing deeper as the words of another song came plainly from the house : —

" The miller he lived close by the mill,
    And the wheel went round without his will.
    With a hand in the hopper and one in the bag,
    As the wheel goes round he cries out, 'Grab!'"

Evidently the words of the miller were taken
literally, and the scurrying about the room and the
shrieks of laughter that followed clearly showed
the actions of the people within the building. The
smile of sympathy on the face of the approaching
man deepened, and, unmindful of the fact that
there was no one to hear his words, he said aloud :
" I heerd there was t' be a sugarin'-off party at
the Curtises' t'-day. Guess all the young folks o'
the neighborhood has got together, if the racket
they're makin' is any sign. Wonder if I'll hev
time t' stop and get a bit o' sugar?" he added in-
quiringly, at the same time glancing at the west,
where the sun had already disappeared from sight.
A dubious shake of his head followed the look, but
he nevertheless called sharply to his steed and

loosely slapped him with the reins. The gait of the faithful animal was unchanged, however, by the encouragement, and the monotonous sounds of his jog-trot steadily continued.

At last, as the "post" man arrived in front of the house, which stood not far back from the roadway, he flung the lines over the back of his horse and slowly descended from his seat in the sleigh. Then taking the one letter which he was to deliver, he slowly approached the door, still more deeply interested in the sounds to which he was listening than in the delivery of the missive in his hand. Only once had he glanced at the directions inscribed upon it, a fact that would have assured all who knew him that Enos Baker must certainly be absent-minded or intent on other matters, for it had come to be a part of his bounden duty, as he regarded it, not only to deliver the epistles intrusted to his care and collect the amount due for such delivery, but also to ascertain, if possible, what was contained in the letters and convey such choice bits of gossip or information as he secured to his wife, who in turn was not slothful in scattering the news, whatever it might be, among her friends in the sparsely settled region.

"Hi! Here's Enos!" "Come in, Enos!" "Give us the letter!" "Come join us, Enos!" "What news, Enos?" The door had suddenly been flung open as the postman approached, and in the door-

way stood a group of laughing young people eager
to hail the coming of the postman, for a letter was
an event of no light moment in those spring days of
1757. Enos, delighted at the greeting he received,
held aloft the letter he was to deliver, and then
said slowly as he drawlingly read the direction : —

"It seems to be for Peter Van de Bogert. I was
told that I would find him here. His aunt said
there was to be a sugarin'-off here t'-day, and that
Peter would be here. But I don't see him."

"Here I am, Enos," responded a sturdy young
man, approaching from the rear of the group.
"Give me the letter."

"Let me see the color of your sixpence, first."

"Sixpence? I have not a farthing with me.
Give me the letter, and the next time you stop at
my aunt's house you shall have your money."

"Nay, it is sixpence before, not sixpence after."

"But I tell you I have no sixpence here."

"Then the letter goes back into my pocket."

"Shame, Enos!" cried one of the girls, for there
were as many of them in the little company as of
their bolder companions. "Give Peter his letter!"

"So I will, Sarah, when he gives me my sixpence."

"He will give it to you. Peter always does
what he says he'll do."

"I have more promises now than I know what
to do with," responded Enos, doggedly.

The door had meanwhile been closed behind him, and suddenly Sarah began to sing : —

> "With a hand in the hopper and one in the bag,
>   As the wheel goes round he cries out 'Grab!'"

As if by some prearranged plan, when the word "grab" was shouted, all the half-dozen girls present rushed upon the unsuspecting postman, and as he was pulled about the room, the shouts and laughter being redoubled, suddenly Enos lost his grasp upon the epistle and angrily faced the assembly.

"A parcel of hussies!" he shouted. "Things have come to a pretty pass when girls act like this. When I was young they were taught to know their manners!"

"There, Enos, we meant no harm," said Sarah, soothingly. "You must taste our sugar before you part from us. I'll get you some now." And in a brief time the energetic and laughing girl placed a wooden dish before the postman. "It's only a little we could make, Enos, you must know, for the sap is not running much as yet. We must have more sunshine for that; but we wanted enough so that we might show the boys who are going away soon that we could have a parting sugaring-off, and we've made just about enough for that. And you shall have more than your share, Enos, for being so good as to give Peter his letter. It was good of you, Enos, to give it," she added demurely.

"I like not such ways," muttered the post-man, nevertheless reaching forth his hand for the tempting dish. "And I have not the six-pence."

"Here is your sixpence," said Sarah, glibly. "Peter shall pay me."

"You don't know how much I have chalked up on my barn door," said Enos, as he placed the coin in his wallet. "It's so much I have to be careful how I add to it. Of course I know Peter would pay —"

"What's the news of the Frenchmen, Enos?" interrupted one of the young men.

The postman shook his head slowly as he replied, "No good news."

"You didn't expect 'good' news of the French-men, did you, Enos?" inquired Sarah, sharply. "I never heard of a good one yet, myself."

"You're not as old as you will be some day, Sarah," said the postman, soberly.

"I doubt not that when I am as old as you are, Enos, I shall have seen many things; but I never expect to see a good Frenchman."

"I'm not so old as ye're thinking, my lady!"

"Yes, you are. You must be, Enos."

"How old d'ye think I be?" demanded Enos, tartly.

"Oh, you must be forty."

"Humph!" retorted Enos, mollified at once. "I'll

tell my wife o' that when I get home.   Yes, I'm
forty.   I'll own up to that."

"Enos, you've been forty for the past forty
years, haven't you?" said Peter Van de Bogert,
soberly.

A laugh followed the rude inquiry, which caused
the face of the worthy postman to flush with
anger; but Sarah, quick to perceive that the old
man's feelings had been hurt, at once urged upon
him the duty of eating the maple sugar while it
was warm, and soon the smile of satisfaction again
appeared upon his weather-beaten countenance.
The rude and boisterous games, however, were dis-
continued, and frequent glances were cast at Peter,
who, at one of the little windows, was industri-
ously striving to read the letter he had received.
His round face betrayed nothing to the young
people watching him, and when at last he finished
his task and thrust the letter into his pocket, it
was Sarah who said:—

"Well, what news, Peter?   You know we all
want to hear, for a letter is no common event with
us."

Peter smiled as he replied soberly; "It will not
break up our party, anyway.   It has no bad
news."

"Is it about the Frenchmen?   Do you have to
go back to Fort William Henry?   Why don't you
tell us?"

"You seem to think everything concerns the Frenchmen, Sarah."

"So I do, so I do. I can't, for my part, see what all the trouble's about, anyway. Isn't there room enough here for them and for us, too?"

"No, there isn't, Sarah!" spoke up Enos, quickly. "There isn't room enough for them anywhere on the earth. They're a bad lot — always were and always will be."

"But they made settlements over here as well as the English and the Dutch, didn't they?" persisted the girl, speaking as much from a spirit of mischief as from a conviction that what she was saying was the truth.

"They claim enough, there's no doubt about that," admitted Enos; "but that doesn't mean that they own all they claim, does it?"

"I have heard they have sixty forts in America. That looks to me as if they thought they had something to protect, anyway!" [1]

"Their whole population isn't more'n a hundred thousand." [2]

---

[1] In 1750 the French had a chain of sixty forts in America, extending from Montreal to New Orleans. On or near the sites of these old forts are many modern cities, such as New Orleans, Natchez, Vincennes, Fort Wayne, Toledo, Detroit, Peoria, Mackinaw, Montreal, Ogdensburg, etc.

[2] Within the present confines of the United States there were probably not more than seventy-five hundred Frenchmen, the most of whom were fur-traders or adventurers.

"How many have we in our colonies?"

"'Bout a million and a half."

"Then what's all this war about? If we have so many and they have so few, why don't we just stop them, or drive them out?"

"That's what we're goin' t' do," said Enos, sharply. "But ye mustn't forget that the French-men have got the redskins in the muss, too."

"I'm not likely to forget," said Sarah, soberly, and a hush fell over the entire company at the reference to the red-skinned allies of the French. "Here's Peter, he knows all about it, for he has been with Sir William Johnson and can tell us just what they do," she added, looking at Peter as she spoke.

"I don't know much about it," responded Peter, quietly, "but what little I do know is more than I wish I did. Very likely I'll know more soon."

"I must be goin'," said Enos, rising from his seat, "or I'll be seein' some o' 'em myself afore I get home."

As he finished his preparations for departure and approached the door, it was suddenly opened from without and a man stepped into the room. He was a hunter or trapper, as his manner and garb would have proclaimed had he not been personally known by most of those in the room. His unex-pected entrance had instantly put a stop to con-versation, and the eyes of all were fixed upon him.

Tallow dips, or candles, had been lighted by this time, and the weird appearance of the man was more marked in the dim light. But apparently ignoring all others, the newcomer at once advanced toward the place where Peter Van de Bogert was standing and called him by name.

CHAPTER II

The Postman's Snakes

" I'VE come for you, Peter," said the hunter.

"So I thought," responded Peter Van de Bogert. " I have just received a letter in which I was told that I must go back to the fort at once."

"I don't know anything 'bout 'letters'!" retorted the trapper. " I can't make out any o' those funny little quirks, that are worse than the tracks o' ducks' feet in the mud. When I have anything t' say I most generally always put straight for it myself, and say it an' have done with it. That's why I came for you. Your aunt told me you were here."

"Well, here I am, Sam."

" So I see, so I see, an' glad I am to find ye, too."

The hunter's face beamed genially, and his satisfaction at meeting his young friend again was apparent to all in the room. Nearly two years before this time[1] both men had been with Sir William Johnson and had had a share in the stirring events which had culminated in the battle of Lake George, where sturdy Colonel Ephraim

[1] See " With Flintlock and Fife."

Williams had lost his life. In the year which had intervened after that campaign, the English colonists had been busily engaged in erecting and strengthening Fort William Henry, while their French adversaries had been equally busy in building at Ticonderoga a strong fort which they had named Fort Carillon. Hither, from Montreal, had come soldiers and supplies, and though no open engagement had occurred between the opposing forces, nevertheless many exciting encounters had taken place between the men and the Indian allies of the two armies.

In these the hunter and trapper, Sam, had had more perhaps than his due share. Erect, restless, impatient of discipline because of his life of freedom in the woods, he nevertheless, both because of his skill and his courage, had been of great assistance to the colonials, and now that the struggle between Great Britain and France was likely to become more active in the New World as well as in the Old, his heart was eager for the coming fray. In his prejudice against many of the Indians, or "redskins," as he termed them, he had come to look upon them much in the light with which the impulsive girl, Sarah Curtis, regarded them, and his feeling toward the Frenchmen, also in part due to his prejudice as much as to his principles or convictions, was likewise exceedingly strong.

His young friend, Peter Van de Bogert, twenty years of age, round-faced, strong-bodied, trained to all the rude experiences of frontier life, with no living parents and only the name of a home with his aunt in one of the little settlements, had been more than willing to enlist in the struggle, although his feelings were neither so intense nor so strong as those of his older companion. For several months past he had been with the little garrison at Fort William Henry, and the short furlough granted him had been the occasion of a brief visit home, where the "party" already described was given by his friend Sarah in his honor. The ten days promised him had not yet elapsed, but the letter which Enos had just delivered was from his captain, ordering him to return to the fort at once, and the word had been supplemented by the unexpected arrival and summons of the hunter, who was universally known in the settlements and in the forts as "Sam."

The coming of Sam had caused Enos to hesitate over his own departure from the house, for his feeling of curiosity was keen, and ofttimes overmastered his good judgment or even his sense of duty. Like all who "hesitate," he was speedily "lost," and reëntered the room, where he was soon listening in open-mouthed wonder to the marvellous tales of Sam's own exploits, which he was giving in response to the eager questions of the young people in the assembly.

"But what I've been tellin' ye 'bout," Sam said, when many of his numerous stories had been related, "isn't anything t' what ye'll all hear of before the leaves drop from the trees this season, I'm thinkin'."

"Sho! Don't ye be too certain sure o' that, Sam," protested Enos, who did not take kindly to the monopoly of the attention and interest which the hunter had secured. "I don't waste any love on the Frenchmen, and I shouldn't be at all surprised to see 'em leave before a gun was fired."

"You're not very easily surprised, Enos," laughed Sam; "but if you'll come along with Peter an' me, ye may see somethin' to open yer eyes more'n that rattler did that I killed for ye last summer. Remember that, Enos?"

A laugh arose from the young people, for Enos's stories of his dealings with snakes made one of the current jokes of the region. The face of the postman flushed angrily at the reference; but restraining his impulse to retort sharply, he said, as he smiled foolishly: "Oh, well, ye'll have yer joke. an' I can't stop it, I s'pose. 'Twas a rattler ye killed, Sam, an' ye mustn't forget that I never said a good word for them, any more'n I did for the Frenchmen. Rattlers is the Frenchmen and redskins 'mong snakes."

"What kind is it you raise, Enos?" inquired Sarah, mischievously, aware of the smile of understanding that at once appeared on the faces of her companions.

"'Twasn't rattlers," asserted Enos, doggedly.

"What was it?" repeated Sarah.

"'Twas black snakes; that's what 'twas!" retorted Enos, sharply. "There's as much difference between a black snake and a rattler as there is between Peter Van de Bogert an' a Frenchman or a redskin."

"But you do raise them, don't you?" persisted Sarah.

"No, I don't 'raise' 'em! I jest had a pair o' 'em last year, an' they were better friends, let me tell you, than some folks is. I'd rather have a good black snake any time than a dog. I mean jest what I'm sayin'!" he asserted boldly, as a laugh again arose from the assembly.

"I heard that you had a great time with them, Enos," said one of the young men. "What was it you trained them to do?" As he spoke the young man glanced slyly about him, assured that in referring to what was a well-known foible of the worthy postman, he had touched upon a matter of interest to all concerned.

"Oh, I don't mind tellin' ye some o' the things they did. I know there's some folks what don't like snakes, an' what pretend for to hate 'em, but that's jest because they don't know 'em. If they knew 'em, they wouldn't feel that way. Now last

summer we had a pair o' black snakes in our barn,
an' they was just as affectionate and good as
dogs."

"How many did you say, Enos?" inquired Sarah,
who was in the centre of the group that now
surrounded the postman.

"Two.   Their names was Ant'ny an' Cleopatry.
I didn't ever think o' givin' names t' 'em, but a
man come along one day by our place an' he was
so int'rested in 'em that he said that was what
we ought to call 'em, an' so we did.   Good names
they was, too!" he asserted boldly, as if he were
fearful that some one would dispute the claim.

"Of course they were good names," asserted
Sarah.   "But what was it you trained them to
do?   You haven't told us about that."

"That's jest what I'm tellin' ye.   I know it
sounds pretty large, but what I say is the truth an'
nothin' but the truth.   My wife M'hitibel, as everybody knows, is a tender-hearted woman, and when
she kills a hen she feels so bad she jest has to put
a cloth around her head before she cuts it off."

"Does she cut off her head often?" inquired
Sarah, soberly.

"You *know* I'm referrin' to the hen's head," said
Enos, ignoring the laugh that arose at Sarah's
question.   "But those two snakes got so trained
that they pretty soon saved M'hitibel's feelin's
perfectly."

"How did they do it?" inquired Sarah.

"Why, they seemed to know just what hen 'twas that my wife wanted to kill, an' one o' 'em would just creep up and crawl 'round th' neck o' that hen, an' afore ye knew it, it would have the hen all stretched out dead. It would just fasten itself round the hen's neck an' squeeze it, an' that's all there was to it. M'hitibel didn't b'lieve the hen suffered a mite."

"Was it Ant'ny or Cleopatra that usually did it?" said Sarah.

"Sometimes one an' sometimes th' other."

"I'm thinkin' if your wife hadn't been there the hen would 'a' been dead, jest the same," suggested Sam; "an' if she hadn't appeared on the scene pretty sudden, there wouldn't 'a' been much left o' that hen."

"Yes, that's what some folks has said; but how d' those snakes know jest which hen 'twas she wanted killed? Tell me that, will ye?"

"I can't," said Sam, soberly. "But it might be that instead o' their killin' just the one she wanted, she always wanted just the one they'd killed."

"No such thing! But they did more curious things than that. My little boy, Mephibosheth, he's a bit lame, that's why we give him his name, used to go out in the summer-time to pick black-berries — "

"Nothing strange about that, Enos. Anybody

can pick blackberries in the summer-time. Now if he'd gone out in the winter to pick them, that would be something wonderful," suggested Sarah, laughingly.

"Just you wait. These two snakes would go along with him, and he'd just watch 'em; an' pretty quick they'd coil up in some bushes, an' right there Mephibosheth would find the sweetest berries ye ever tasted."

"Nothing queer in that, either," said Sam. "Everybody knows that black snakes 'll eat blackberries, and all your boy had t' do was just watch 'em. Course they'd find the best!"

"That's right, Enos. Sam has explained it. There's nothing strange in what you've been telling us," said one of the young men, calmly, although he was aware of the interest and delight of his companions. Enos's stories were proverbial in the region, and only a slight incentive was required to make him increase them in number and in size.

"Well, here's one thing they did, anyway, and ye can't explain that away. My wife M'hitibel does the milkin', an' sometimes she gets the rheumatiz so t' she jest can't move. We don't keep but one cow —"

"And of course you do the milking when your wife is sick," interrupted Sarah.

"Can't say t' I do," replied Enos. "I never was any good at that job, an' besides I'm so busy I can't be depended upon. No, if the cow didn't get milked till I milked her, I'm afraid she'd go dry. Well, this time I'm tellin' ye 'bout, M'hitibel had the rheumatiz dreadful, an' she couldn't get out no-how. Now those two snakes seemed t' understand jest 's if they'd been humans, an' the first thing ye knew they'd gone off an' got a couple o' milk snakes, an' when my wife at last managed to crawl out t' the barn, there was the cow all milked. Yes, sir, jest milked as clean as M'hitibel could 'a' done it."

"Did they milk into the milk-pail?" inquired Sam.

"Partly, partly. Course some o' the milk was lost, but not all."

A shout arose at the postman's words, and glancing angrily about him he seized his skin cap and made for the door.

"Hold on, Enos," called one of the young men. "Don't go yet. There must have been something else those snakes did. Tell us one more. What was it?"

"No use in tellin' ye. Ye act, every one o' ye, as if ye didn't believe a word I said."

"Please, Enos. There must have been something more. They were the most wonderful snakes I ever heard of," said Sarah.

"They was. They saved my life once."

"When? How did they do it?"

"Well, ye see, as I was tellin' ye, I never liked rattlers. One day last summer I was out in our pastur', an' all at once a big rattler just rose up behind the rock and looked straight at me, an' 'twasn't a minute 'fore he charmed me. I could see more'n a million rainbows as I looked into his eyes, an' of all the music ye ever heard in all your born days! There was fiddles, an' a drum, an' a fife, an' — an' — I don't know what all. But 'twas fine, every bit o' it. I knew 'twas all day with me, but I couldn't move. Couldn't stir hand nor foot. I jest stood there lookin' into that rattler's eyes, an' listenin' t' his twenty-seven rattles — "

"Did you stop to count them?" interrupted Peter.

"And I knew 'twas good-by for me. The rattler kept a comin' closer and closer, and there I was just as if I'd been a stump. All at once up comes Cleopatry, an' like a flash o' lightnin' she threw herself on that rattler, and before ye could wink yer eye she had the tail half o' him all tied up. She jest twisted herself tight as a drum string 'round him, an' there he was! Then up comes Ant'ny with his mouth wide open, an' quick as a wink he swallowed the whole o' the other half o' the rattler."

Shrieks of laughter interrupted Enos, and the postman again seized his cap and started for the door.

"Come back, Enos!" "Don't leave the snakes there!" "Did they ever get away?" "What became of you?" "Is Ant'ny still where he was?" were among the questions that followed him. But this time Enos was determined and did not once look behind him. In a brief time he had mounted to his seat in the sleigh, and the monotonous gait of his departing steed had been resumed.

CHAPTER III

Into the Forest

THE departure of the postman was the signal for the breaking up of the party. There was the evening work to be done, and not one of the boys or young men who were present was without a task in the recently acquired homes.

"When do you start for Fort William Henry?" inquired Sarah of Peter, when the two chanced to be by themselves for a moment.

"That I cannot say," replied Peter. "'Twill be as Sam says. We may go to-night, and we may not start before to-morrow morning."

"We shall not forget you, Peter."

"That's good of you."

"I hope nothing will happen to you. There's one thing I know will not, whatever else may come to pass."

"What's that?"

"You won't be shot in the back."

"Thank you. I shall try not to let that occur," replied Peter, laughing lightly, although the words of the eager-hearted girl touched him more deeply than he would have acknowledged.

The good-bys were spoken soon afterwards, and the young people departed from the home of Sarah Curtis, few of them realizing that the gathering was the last which some of them would ever attend.

Peter Van de Bogert and the hunter left together, and for a brief time not a word was spoken as they trudged on over the rough roadway. The younger man was the first to break in upon the silence, as he said : —

"Sam, what's ahead ? "

" That's as may be," responded the hunter, enigmatically.

" Anything new at the fort ? "

" Not exactly at the fort."

" What is it, Sam ?   There's some good reason for my letter and for your being here as you are."

" That's a fact, Peter !   No mistake about that! Well, the truth is that the Frenchmen are becoming very lively, and there's no knowing when they may try to strike the fort.   They tell me that this new general they've got is a wonder."

" Montcalm ? "

" Yes, I believe that's what they call him."

" When do we go, Sam ? "

" I'd go now if 'twasn't for the storm.   There'll be snow before mornin', or I'll miss my guess," the hunter added, as he gazed for a moment at the overcast sky.   "If you and I, Peter, were the only ones to go, I wouldn't wait; but — "

"Others going, too? How many, Sam?" interrupted Peter.

"Four, and maybe five. We'd better put up at
your aunt's to-night and start out in the mornin'.
I think we'll save time."

"You know best," responded Peter, quietly. And
not another word was spoken until they arrived at
the log-house of his aunt, where they both entered
and soon passed up the ladder to the loft which
Peter occupied. It was the only home he had ever
known, for, left an orphan when he was only a
babe, he had been not only without any knowledge
of father or mother, but also of brother and sister
as well. It had been a frequent remark of his that
he had never known any one who was so completely
alone in the world as he. Not that he was unwel-
come in the family of his aunt, but he never was
able entirely to overcome the consciousness that,
after all, it was not his home, no matter how
cordial or kind his aunt and her mother, his grandmother, were to him. Perhaps this fact was one
of the reasons why he had been so willing to enter
into the struggle which the colonists were making
to keep their lands from the grasp of the encroaching Frenchmen and their fierce and savage allies.
Those who knew him best were assured that his
serious manner and somewhat sober countenance
— for his round face seldom was lighted by a
smile and he was exceedingly earnest in all that he

attempted to do — were, in a measure at least, due to his knowledge that he was almost alone in the world.

However, when early on the following morning, after having breakfasted by candlelight, he and his friend prepared to depart, the words of the two women were not lacking in tenderness, and Peter's heart glowed whenever he recalled them in the days that followed.

The snow had ceased falling, and the air was crisp and clear. Strapped upon the backs of the two men were the bundles containing the few necessities they were taking with them. Each was equipped with a pair of snow-shoes, and Peter had also taken a pair of skates. These skates were roughly made, and the blades were long and curled at the end where the iron was turned back over the foot of the skater. The hunter at first had laughingly objected to his companions burdening himself with the additional load, but perceiving that Peter was not to be turned from his purpose, he said : —

"You won't have much time for using those things, I'm thinkin'. It'll be all work and no play when we get inside the fort."

"I am taking them to work with," responded Peter, quietly. "I may not want them, and then again, I may."

"Yes, that's as may be," assented Sam, sagely.

They were soon joined by their four companions, and at once the little band set forth on their journey. The settlement was soon lost to sight, and, led by the hunter, they soon abandoned the roads to plunge into the forest, making their way with sturdy strides amongst the giant trees. The recently fallen snow, light and feathery, wrapped the long branches in its mantle and covered the valleys and hills in a robe of white. The silence of the party seemed almost like a reflection of the stillness that rested over the wilderness itself. For mile after mile they pushed forward on their way, moving in single file and following the hunter, who, by common consent, had been made the leader of the band. At noon time they halted in the shelter of a small valley and kindled a fire, over which they cooked the food they had brought with them. But only a brief time for rest was taken, and they speedily resumed their journey, going steadily forward until the sun sank low in the western sky.

At the word of the leader a halt was then made, and soon boughs of young cedar trees were cut and a shelter was made for the night. It was deemed safe, as they were not yet in the region where the presence of the enemy was to be feared, to have a fire, so in a short time a great pile of logs was heaped up and a roaring fire blazed in front of their sleeping-place. When they had eaten their supper, the men wrapped themselves in their blankets and were speedily sleeping, the precaution of stationing some one of their number as guard not being thought necessary.

Peter was the last of the little band to close his eyes. The weirdness of the place, the whistling of the wind among the naked branches of the great trees, the roar of the flames and the light that leaped from the fire, combined with the recollections of his brief visit in the settlement to make him wakeful. Again he lived over the events of the past few days, and the games and mad escapades of the "sugaring-off" at the home of Sarah Curtis were recalled. When would he again be with the young people he had known since his earliest childhood? That stirring events were before him he was well assured, for at last the two nations had roused themselves for a supreme test of their strength. Whatever might be the results of the struggle in the Old World, in the New, at least, it meant more than the mere ability to plant the lilies of France above the banner of the lion and the unicorn. It was a struggle for home, for the protection of women and children, a defence against the merciless Indian, jealous of his own rights to the region and ever quick to resent the encroachment of the white men into the country where his forefathers had roamed at their will. And now had come the supreme contest from which there was to be no turning back.

All of these things revolved in the mind of the young soldier, but at last the roar of the flames and the sighing of the March wind blended strangely in his ears and soon ceased to be heard altogether, as, wearied with the day's march, he fell asleep.

On the following morning the journey was resumed, and the day passed much like that which had preceded it. At nightfall another halt was made, but this time, although a protecting shelter was built of cedar boughs, it was not thought wise to have a fire, for the region was now at hand where lurked the peril from prowling Indians or where scouting parties of the hated Frenchmen might be met.

The night slipped by without event, and at daybreak, when the band was once more astir, Sam, who had been preparing the simple breakfast, joined his comrades, and after a brief silence said : —

"I've a notion that we'd do well to change our plan now. We ought to be inside the fort by sundown."

"What is it you want us to do, Sam?" inquired Peter, looking up quickly, as the hunter spoke.

"I've been thinkin' of it ever since I got up," replied Sam, "an' I've thought it might be a good thing for us to spread out an' no two of us come into Fort William Henry together. Ye see," he added, as he perceived the unspoken protest of some of the men, "the Frenchmen are gettin' to be mighty pert, an' some o' 'em has been seen right close up to the fort itself. Now if we all spread out somethin' like a turkey's tail, an' come into the fort from all sides, we'd be likely t' find it out if the' were any parties lyin' 'round here from Ti."

"Yes, but they might find us, too, Sam," suggested one of the party.

"That's as may be. Every man o' us has got two eyes an' two ears, an' he can keep 'em all open, can't he? If he walks into a trap that's spread out for him right afore his nose, he might as well give up then 's any time, for I'm thinkin' he wouldn't be much good to the fort if he were inside; so he might 's well be out 's in."

"We might lose our way," suggested another.

"I guess not, though that's as may be, too. Have t' take some chances when ye're fightin' redskins and Frenchmen, same 's ye do when ye set your traps. But I don't believe ye'll have much trouble if ye follow my d'rections careful."

"What is it you want us to do, Sam?" said Peter, quietly.

"Jest what I said. We can spread out and come into the fort from all sides, and if there's any sign o' the enemy bein' hereabouts, we'll be apt to see it."

"Do you mean for some of us to go to the other

side of the fort and come down to it from that
side ? ”

“ Exactly.”

“ Who is to go there ?   Do you want me ? ”

“ Yes, ’s far ’s I’m concerned, that’s just what
I want.”

After a brief discussion and some demurring on
the part of a few, it was decided to adopt the
hunter’s suggestion that, instead of keeping together, as they had been doing, they should now
separate and every man strive to make his way
alone into the fort from the direction to which he
might be assigned.   It was finally agreed that the
hunter and Peter should go some distance to the
farther side of the fort and approach it from that
quarter, while their four comrades, spreading out
after Sam’s suggestion “ like the tail of a turkey,”
should strive to enter from the side on which they
then were.   Not many miles of their journey now
remained, and it was confidently believed that by
nightfall, or soon after, all six would have gained
the protection of the fort.   Certainly the four men
who were not to make the longer journey, or enter
the more perilous region, ought easily to be able
to accomplish this, and with this assurance the
band soon broke up and the journey was resumed.

Until noon Peter and the hunter kept together,
but after they had eaten their hasty meal, they too
separated, and each departed on his own way.

Left to himself, Peter Van de Bogert began to realize more completely the loneliness of his position. If prowling enemies were about, as according to reports they assuredly were, he must look to himself alone for protection. No one had been seen by him when late in the afternoon he arrived at the shore of the lake, at the place which had been assigned to him, but then there came a sudden change in his plans and actions as well.

CHAPTER IV

A Race on Skates

A S Peter, standing on the shore, gazed out over the lake before him, he perceived that the light snow had been blown about by the March winds, and that for the most part the ice was clear. It was true that in places he could see small drifts, and that the ice itself was not uniformly smooth; but it was now possible, he decided, for him to abandon the use of his snow-shoes and make his way to the fort by means of his skates. At all events it was worth a trial, and he quickly removed the snow-shoes, strapped them to the bundle he had been carrying upon his back, adjusted his skates, and prepared to depart.

According to his reckoning, Fort William Henry was not more than five or six miles distant. He was aware that his peril might be increased by moving openly over the ice instead of continuing on his way among the trees that grew close to the shore, for now his movements could be readily seen by any party that might be concealed near by. But the very nearness of the fort and his own skilfulness as a skater combined to induce him to accept the added risk, and in a brief time he was gliding swiftly along, keeping well in toward the shore, and watchful of the region he was approaching.

The sun was hidden from sight by the cold gray clouds banked in the western sky, and the wind blew strongly in his face, as with long, steady strokes he swept forward. There was a loneliness in the apparently deserted region, for not one living object could he see. The noise of his skates became monotonous, and an indefinable dread impelled him to redouble his efforts. His speed increased, and despite the cutting wind his head was soon wet with perspiration. On and on he urged his way, the longing to gain the protection of the fort and the ill-defined but nervous fear continually providing a fresh incentive.

He had gone what he thought must be a full mile and a half when he suddenly checked himself and swerved sharply from his course. From a little wooded point projecting before him he saw a half-dozen or more men suddenly emerge and spread out before him, directly across his path. It was not light enough to enable him to see who or what they were, but their actions were sufficient to convince him that they were in no wise friendly. Besides, had not the hunter told him that prowling bands of the French and Indians had recently crept close up to the walls of Fort William Henry?

At first a wild impulse to make for the more

open part of the lake and attempt to pass the men
by a burst of speed seized upon him; but in a
moment he changed his plan, when he perceived
that some of the party were also equipped with
skates, as he was, and that their actions betrayed
the skill with which they were able to use them.
There was nothing for it but to turn back, and as
the men were still at a considerable distance from
him, he might be able to gain the shelter of the
adjacent forest, and in the darkness, which soon
would cover all, throw them from the pursuit and
make his way to the fort by land.

Already he had turned, and with redoubled
speed was striving to gain the shore of a small
bay or cove that he could see not far away. He
glanced hastily behind him and saw that, swiftly
as he himself was moving, his pursuers were equally
fleet, and as they were not encumbered with any
bundles such as he was bearing upon his back, if
the chase should prove to be a long one, there could
be no doubt of its final outcome. Yet the thought
only nerved him to greater endeavor. The keen
wind cut sharply into his face and retarded his
progress somewhat, but as he realized that his
pursuers must face the same conditions, his determination to escape increased. His weariness was
for the time ignored or forgotten, and after the first
fear had passed there came a feeling almost of
exultation. He would gain the point he desired,

he assured himself, long before his pursuers could overtake him, and once in the woods his snow-shoes would afford him the means of soon leaving his enemies far behind him. In the confidence that now possessed him he glanced again behind him and tauntingly shouted his defiance.

An answering shout rose on the air, but Peter Van de Bogert's heart almost stopped for the moment when he realized that the cry came, not from the men behind him, but from some who were before. Instantly he turned and glanced in the direction from which the hail had been heard, and to his consternation he perceived that four men had moved out upon the ice from the very point which he was striving to gain.

With a sinking of the heart the young soldier instantly realized that he was between two lines of his enemies, and what was even worse was the fact that the men before him, as well as those behind, were provided with skates, and evidently knew how to use them as well as he himself did.

For one brief moment Peter was tempted to make for the immediate shore, but the men before him could cut him off before he could gain it, and the plan was instantly abandoned. He quickly decided that his only hope lay in making straight for the open lake and, perhaps, for the opposite shore. His own course would be direct, while both parties of his enemies would be compelled to go a greater distance than he, as one moved from a distance behind him and the other was almost as far before.

There was no time for confusion or hesitation now, and exerting all his strength, Peter started directly for the open lake. A cheer rose from the band behind him as his action was seen, and a responsive cheer came from the men on the other side. It was evident that they thought they had their game almost as good as caught now, and would soon be able to run him down. But Peter was not yet disheartened, although he fully realized how desperate his predicament was.

His strides were longer now and he was endeavoring to make every moment tell. He had, in his younger days, frequently entered into a contest with his boy companions, to see who could make the straightest line of tracks across some field where the snow had recently fallen, and he recalled the words of Sam, who one day had been an interested observer of the sport. He had told him never to look at his feet or the tracks he was making, but to fix his gaze upon some object on the side of the field he was seeking and move straight for that. Even now, in his excitement, Peter recalled not only the directions which the hunter had given, but the success which had crowned his own efforts the very first time he had adopted his friend's advice.

In his present emergency, when the prize might

be safety, if not life itself, Peter endeavored to be
calm and to follow the suggestion which had won
for him success in his boyish contests.   Far away
he could see the dim outlines of a small island or
part of the mainland, he could not determine which,
and looking neither to the right nor left he made
it the goal for which he was striving.

For a time the race continued without interrup-
tion.   He had sharpened the runners of his skates
before leaving home, and the speed at which he was
moving proved the wisdom, as well as the success,
of his well-timed preparations.   Long lines and
deep were cut in the ice as he sped forward.   His
strokes were even and powerful, and he was moving, he felt assured, as he never had before.   The
object far away, upon which his gaze was fixed,
was beginning to assume more definite outlines.
His present rate of speed must bring him to its
shelter before his pursuers could overtake him he
felt certain, and in his confidence he glanced, for
the first time, at the men above him.   Assuredly
they had not decreased the distance between themselves and him, and, with a feeling of exultation, he
turned to look at the men below.

Suddenly his skates struck a mass of ice where
the snow had collected, melted, then frozen again,
and its rough and projecting points for the moment almost destroyed his balance.   His arms
waved wildly; it required all his skill to prevent

himself from falling. His violent efforts loosened
the thongs of deerhide on his right skate, and in an
instant it was freed from his foot and went speed-
ing over the ice to one side, while he himself was
thrown and began to slide in another direction.
Bruised and almost stunned by the fall, he nevertheless clutched wildly at the rough ice, striving
desperately to check his speed, but despite his efforts he was borne thirty feet away before he could
rise again.

He was aware of the shouts that greeted his mishap, but in his desperation he gave them no heed
as he started toward the place where he thought
his missing skate must be. In the dim light it
seemed to Peter that hours had elapsed before he
found it, yet, despite his eagerness, he carefully re-
tied the thongs and made positive that they would
not become loose again, before he resumed his flight.

His pursuers were nearer now, and their shouts,
distinctly heard, indicated that they were confident
of capturing him. That this was what they were
striving to do Peter felt assured, for otherwise long
before this time he would have heard the reports of
their guns.

Again the race was resumed, and soon Peter
realized that the violence of his exertions was beginning to tell visibly upon him. His breathing
was becoming labored and there was a dull pain in
the side upon which he had fallen. But not for a

moment did he relax his efforts. He was dimly aware that on either side some of his pursuers were gaining and already were abreast of him. It was plainly their purpose to gain a place in advance and then turn back and catch him in their midst. Only slightly decreasing his speed, Peter unfastened the straps by which the bundle upon his back was bound, and as it fell upon the ice, relieving him of the load, he could see for a brief time that he was moving a little more swiftly. But even this relief did not enable him to pull away from the men who, not more than twenty yards away on either side, were doggedly holding to their course. He could even perceive that they were steadily gaining, and already were slightly in advance. He groaned as he thought of the long and wearisome tramp which for three days he had kept up through the forests. If he was only as fresh as his pursuers, he would have no fear of being overtaken; but the thought brought him no relief, for there was no escaping the fact that he was not equal to the demands upon him, and that in all probability he would soon be in the hands of his enemies.

Suddenly it flashed into his mind that the men who were now a little ahead of him were doubtless the best skaters in either band. The men who were behind might be distanced, if in some way he could meet them alone.

Instantly and almost imperceptibly he slackened his efforts, and although he was still moving swiftly, he permitted the men in front to increase their lead. It was evident that they were now converging and soon would turn back to hem him in.

Suddenly Peter sharply checked his speed and darted back over the way he had come. He was well-nigh hopeless of escaping, but in his desperation it was his last resort. A sharp cry greeted his unexpected action; but aware of his own purpose, he had gained a slight advantage before the others could turn and follow. As he glanced about him he could see that he had left all his pursuers behind except two, and these, one from either side, were moving swiftly toward him. The man upon his right was easily dodged, and then Peter turned and deliberately ran full into the one who was advancing from the left. The man, unprepared for the shock, was sent sprawling upon the ice, and Peter, after a momentary effort to keep his balance, darted ahead with all his might. His pursuers were now all in his rear. The darkness had deepened, but the shore from which he had started was far away. Shouts and calls for him to stop were unheeded.

"Good! Keep it up! The more breath you waste in that way the better it will be for me," thought Peter, as he endeavored to increase his speed. The calls, however, were followed by the sharp report of a gun, and Peter could hear the bullet as it whistled over his head. Capture was not, then, the sole purpose in the minds of his pur suers, and as he realized how desperate was his predicament, he was almost tempted to stop and give himself up.

His efforts, however, were not relaxed, and he still swept on with the fleetness of the wind. He had arrived at the uttermost limit of his powers. This one final attempt to break away was to be his last, he thought, and then, come what might, he could do no more. He was bending low now, moving almost mechanically. It was all becoming unreal and meaningless, and yet he kept on and on, until suddenly he skated on the thin ice that had formed over one of the numerous open spaces in the waters of the lake. There was a snapping and crackling, a loud, wild cry that sounded weirdly on the night air, and Peter Van de Bogert fell headlong into the icy waters.

CHAPTER V

FRIENDS AND ENEMIES

BREATHLESS, Peter rose to the surface and was aware first of all that the sudden fear which had seized upon him that he might be carried under the firmer ice and so be held in the cold waters was not fulfilled. His hands already were grasping the edge, and almost without his knowledge a wild cry for help escaped his lips.

Before he could recover his breath, however, he felt the plunge of another body in the water, and then suddenly he was seized by the legs, torn from his grasp on the border of the firmer ice, and sank once more beneath the waters. It was a struggle now, not so much for air, as to free himself from the deathlike grip in which he was held. The water surged and roared in his ears. Strange lights danced before his eyes, and he felt as if he were held in by iron bands. At last he managed to strike the man who was holding him, but the firm grip was not relaxed, and in despair he realized that apparently there was to be no escape. The roaring ceased, the dancing lights no longer appeared, and then, although Peter knew but dimly what was occurring, he and his companion were both drawn to the surface and dragged out upon the firm ice. There they were stretched upon coats that had been spread for them, and their arms and legs were rubbed briskly by not unfriendly hands. In a brief time the young soldier perceived his surroundings and remembered what had befallen him. His teeth were chattering and his body trembling as with an ague. It was too dark to permit him to see who the men were that had surrounded him, but the voice of one sounded strangely familiar in his ears.

" You're all right now, lad ? " the man said.

" Y-e-e-s.   Is that you, Sam ? "

" That's as may be, but as far as I can rec'lect it seems to be."

" Where am I ?   What are you doing here, Sam ? "

" Ye're with yer friends, that's who ye're with; though judgin' from yer actions ye didn't seem over glad to see 'em."

" My friends ? "

" That's what I said.   What made you run from us, Peter ? "

" What made you chase me ? "

" We didn't know jest who ye might be, lad. 'Twas dark —"

" And all the time I was trying to get away from our men ? "

"That's as sure 's ye're born, lad. An' ye led us a lively chase o' it, too. I wish ye'd left yer skates at home as I told ye to do."

"Next time I will."

"Came mighty near not bein' any 'next' time to it, lad," said the trapper, more gently. "Think ye're strong enough to go ashore with us now?"

"Ye-e-s," chattered Peter, "I-I-m a-all r-right now."

"Sounds so," grunted Sam. "How's the other fellow coming on?" he added, glancing as he spoke at the prostrate form of the man who had followed Peter into the water.

"He's comin' 'round," responded one of the men, "an' will be all right in a minute or two. You take your man an' go ashore, Sam, an' we'll come along in a little while."

"Think ye can stand it, Peter?" inquired Sam.

"Ye-e-s. I'm a-all right," responded Peter again, "It 'll help keep me warm if I'm moving."

Accordingly Sam and Peter turned toward the distant shore, leaving their companions to follow when the other man should have more fully recovered. Their skates had not been removed in the excitement, and they moved swiftly on in silence, the hunter leading the way and selecting a course of which, apparently, he was certain.

At last, when the border of the lake had been reached, at the word of the hunter their skates

were removed ; but, before they entered the forest,
Peter suddenly said : —

"Sam, I left all my belongings out there on the
ice."

" Ye did ?   What for ? "

" Why, I unstrapped the bundle on my back
when I was trying to get away from you, and I
haven't any idea where it is now.   But I ought to
go back and look it up.   You stay here — "

" You come 'long with me now, that's what ye're
to do ! " responded Sam, sharply.   " Maybe some
one 'll pick up yer belongin's on the way here ; but
whether he does or not, the first thing for you to
do is to get dry."

" But — "

" There isn't any ' but ' about it !   Come 'long ! "

Thus bidden, Peter followed obediently, and in a
brief time found himself at the entrance of a rude
shelter formed of young trees which had been cut
from the surrounding forest.   When he entered he
discovered, in a hole which had been dug in the
ground, the glowing embers of what had been a
large fire.

In surprise he turned to his comrade and said :
" A fire, Sam ?   I didn't think you'd dare have
such a thing here."

" That's as may be.   It's here, anyway, isn't it ?
What you want to do is to get right down close to
it an' dry up an' dry off."

"But why did you have a fire, Sam?"

"Ever hear o' hangin' out a piece of meat to draw a wolf?"

"Yes."

"Well, that's what we might be doin'. Mind ye, I don't say *'twas* what we were doin'. It's jest as may be. Here come the others," he added, as the sound of approaching footsteps was heard.

Nearly a dozen men entered the rude structure, and among them was the one who had gone through the ice with Peter. Mindful of the desperate struggle he had made against him in the water, Peter at once advanced toward him, and extending his hand, said:—

"I beg pardon for striking you. I didn't know just what I was doing. You see, I didn't understand—"

The man glowered at him savagely and refused the offered hand.

Peter's face flushed in the dim light at the rebuff, but striving to appear unmindful, he said: "I'm sorry for what I did. I—"

"Humph!" growled the man. "Ye cracked one o' my ribs."

"I did? I don't see how. I'm sure I did not intend to—"

"When ye ran into me on the ice. What d' ye do that for?"

"I was trying to get away," replied Peter, simply.

"Well, all I can say is that 'ev'ry dog has his day.' Sometime I'll get even with ye. My time's comin' sometime!"

"Here you, Timothy Buffum, quit that!" interrupted the hunter, sharply. "You thought he was a Frenchman, didn't you?"

"Humph!"

"Well, he took you for one, too, so it's six o' one an' half-dozen o' the other. Now that ye know ye're both white men, why don't ye speak up like a man, an' be done with it?"

"He cracked my ribs! He ran into me full tilt."

"Served ye right if he'd cracked yer skull!"

"Oh, I know you, too, Sam!" growled the man, savagely. "You never had anything good to say for me. Ye're agin me, same 's the rest o' 'em."

"Nobody's agin ye, Tim — unless it is yerself," replied Sam, warmly.

"How do I know this man is what he says he is?"

"Ye've got my word for it."

"Who knows that you're all right? You didn't use a shovel at the fort, same 's some o' us did from sun-up till sundown. You were off in the woods an' nobody knows what ye were up to."

"Yes, they did."

"Who?"

"I did. An' what's more, Tim Buffum, if you

want fer t' pick a quarrel with the lad ye'll have t'
reckon with me, too."

"Tim 'll feel better just as soon as he has dried
off," said one of the company, cheerfully. "It
doesn't make a man feel real good to have some
one go at his ribs head first. An' then, Tim's been
in the lake, too."

"Better if he'd stayed there," growled the hunter.

Timothy Buffum's eyes blazed at the hunter's
words and the quarrel threatened to continue, but
some of the men hastily covered him with garments
of their own, and seating him by the glowing coals
endeavored to divert his mind from his injuries, as
well as from those which he fancied he had received
from young Peter Van de Bogert. The man continued to act the part of a churl and refused even to
be seated near Peter, who was deeply chagrined at
the anger he had aroused. He himself had been
cared for by Sam, who, as soon as he was satisfied
that his charge was none the worse for his exciting
experience, prepared to depart, declaring that he
was going to search for the bundle which Peter
had lost in his flight, as no one had found it in the
return to the hut.

"'Twas a lively chase ye gave us," said one of
the men to Peter when Sam had left.

"I did my best; but if I had only known
who you were, all the trouble would have been
saved."

"That's all right," laughed the man, good-naturedly. "I should have done the same myself, if I had been in your place. But tell us what you were doing out there on the lake alone."

"I was on my way to the fort."

A growl from the man who had been Peter's companion in misfortune followed the young soldier's explanation.

"Where from?"

"Back near Albany. I started with Sam. There were six of us, and this morning he thought we'd better spread out and come in from different directions, so that we could report at the fort if we should find anything suspicious."

"Did ye find anything?"

"Nothing."

"I don't know where yer eyes could 'a' been, then."

"Why? Is there anything wrong?" inquired Peter, quickly.

"Plenty of it. That's why we are here, an' why we have ten men besides, out on the lookout."

"How did you find out that I was not a Frenchman?"

"Sam," replied the man, laughing as another growl rose from the unfortunate by the fire. "Ye never see a man more surprised than he was when we hauled ye out o' the water."

"How long had Sam been with you?"

"Oh, he joined us 'bout the middle o' the afternoon."

"Had he seen any signs?"

"Plenty."

"What are you going to do now?"

"That's for the capt'n to say."

"Where is he?"

"Ye'll have to ask him, but he's not very far off,
I guess."

"Are you going to stay here long?"

"Ye're getting all right again, my friend,"
laughed the man. "Ye'd better wait and ask
Sam when he comes back. Maybe he can tell ye
more than I can."

As he spoke the man rose and went out into the
night. Peter could hear that the wind was rising,
and, beginning to feel that he was entirely recovered from his recent mishap, he too got up from
his seat and started toward the exit. He was
compelled to go in front of the man who was
seated by the fire, and as he passed, suddenly a
foot was thrust out, and stumbling over it in the
dim light Peter fell headlong to the ground.

An exclamation of anger came to his lips as he
arose, and the laugh of the man increased his rage.
Turning upon him, he asked sharply, "What's that
for?"

Before any reply could be made, however, the
man who had just departed from the hut came

rushing back, and in an instant all minor things were ignored or forgotten as, breathless with excitement, he began to speak, and at the utterance of his first word his feeling was shared by all who were in the structure.

CHAPTER VI

ALONE

"SAM has come back," panted the man, "and he says that he's certain sure there's a big body o' Frenchmen coming."

"Where?" demanded Peter, instantly as excited as the speaker.

"Over the ice. They're coming up the lake."

"Did he see them? Where is he? What are we to do?"

"Here's the man now," replied the soldier, as the hunter entered the hut. "He can speak for himself."

"No time t' waste in talkin'!" exclaimed Sam. "Ev'ry man wants to get his gun an' be ready for what's comin'. The captain 'll be here in a minute, an' he'll tell us what to do."

As Sam spoke, the leader of the little band entered and turning sharply to the hunter said: "What's this I hear? Out with it, man!"

"The Frenchmen are comin'," replied Sam, quietly. "That's all there is to it."

"How do you know? What did you see?"

"I was out on the ice huntin' for the belongin's

o' my friend," explained Sam, "an' I was certain I
heard the sound o' axes down the lake. I listened
then with both ears, an' pretty soon I could not
only make out the sound o' th' axes, but I could
see fires along near the shore."

"Fires?" demanded the captain, sternly.

"That's what I said."

"What did you make out?"

"I didn't go any nearer, but put straight back
here t' report. But I have my own notions o' what
it meant."

"What do you think it means?"

"It means that the Frenchmen are comin', an'
my 'pinion is that there's a big body o' 'em, too,"
said Sam, positively.

"Why so?"

"Th' fires."

"How do the fires show that?"

"Why, if there was only a few o' 'em, they'd go
skulkin' along an' try to cover their tracks. But
when they stop in the middle o' the night an' build
fires to warm themselves with, why, it stands to
reason that they're not afraid; an' my 'pinion is
that they've got so many men they don't care
whether they're seen or not."

"How far away do you think they were?"

"Two or three miles; not more 'n three, anyway."

"Any Indians?"

"Can't say, but it's nat'ral like to think that

when the Frenchmen are there the redskins will
be, too. Leastwise, that's been th' way o' it afore
this."

For a moment the captain was silent, then he
said sharply : " There's only one thing for us to do.
Some one must go straight to Fort William Henry
and report, and some one must go down the lake
to make sure that what you've seen is true. The
rest of us must stay right here till word comes, and
then be ready for whatever happens. Who'll go to
the fort ? " he demanded quickly.

" I will," said Timothy Buffum. " I'll be glad
to get out o' this."

" Start then ! " and the man at once departed.

" Now I want you to go with me," said the
captain to Sam, " and the rest of you stay right
here where you are. We'll go and find out more
about this, and we'll leave our guards out just as
they are. If anything happens that we're not back
by two hours, then you're to call in the guards and
all put straight for the fort. Now then," he added,
turning to Sam, " we'll start."

The two men at once set forth on their journey,
and as soon as they were gone those who were left
in the hut quickly covered the coals and prepared
to await the return of their comrades. The night
was cold, and although Peter's garments were now
dried, and he had in a measure recovered from the
excitement through which he had passed, the dreary

waiting soon became almost unendurable. The uncertainty as to what lay before them, the conviction
in his mind that what Sam had reported was true,
and the knowledge of some design against the fort
and its little garrison numbering not much above
three hundred and fifty men, all combined to increase Peter's restlessness. When a half-hour had
elapsed, he felt that he could endure the silence no
longer. Several of the men in the hut had already
stretched themselves upon the ground and were
sleeping soundly. How they could so ignore their
peril was something which Peter could not comprehend, but for himself he would endure the
waiting no longer.

Suddenly it flashed into his mind that he had no
gun or weapon of any kind. His own rifle had
been lost when he plunged into the lake, and Sam
had not even told him whether or not he had
brought back his lost bundle. Peter was tempted
to take one of the guns of his sleeping comrades.
He quickly decided, however, that he had no right
to do that, and so, unarmed and alone, he went out
into the darkness.

The hut was located in a thick clump of trees
not more than twenty-five yards from the shore of
the lake, and a brief walk brought Peter to the
place he was seeking. Taking his stand in a secluded spot on the shore, he looked out over the
ice-bound lake. The ice had been swept by the

winds and was nearly clear of snow. The stars twinkled overhead, and the towering hills in the distance were clearly outlined against the sky-line. The very silence made everything doubly impressive, and with the dread which was in his heart, Peter was in a mood to respond to the weirdness of the scene before him.

With bated breath he listened, and peered intently down the shore in the direction from which Sam had declared the noise of the axes had been heard and the light of the fires had been seen. For a time nothing rewarded his endeavors. And yet, somewhere from that direction danger was coming, he felt assured, so strong was his confidence in the hunter's report. Eager in his task, he ignored the passing moments, and did not realize how much time had elapsed since he had taken his lonely stand.

Suddenly he moved abruptly, as a sound came across the ice, dim and indistinct, yet sufficient to stir every sense, already alert with expectation. The sound was soon repeated, and then out of the gloom came the shadowy forms of moving men. The sight was startling, yet fascinating, and Peter watched eagerly from his hiding-place, never once turning away his eyes, and apparently unmindful of any peril to himself. All other things were forgotten for the time as he gazed at the oncoming men. Those in advance had already passed, and behind them came multitudes, as it seemed to Peter. They were marching in regular order, with a sprightliness and spring in their movements that betokened confidence and the absence of all fear. And as Peter watched the moving mass, he was not surprised at their apparent ignoring of any peril that might beset them from the shore, for if there was safety in numbers, then they must be indeed secure. He thought of the little garrison of three hundred and fifty men at the fort, and then, in his mind, contrasted it with the mighty force before him. And still they came, though now the lines were less orderly, and Peter had no difficulty in making out the Indians who brought up the rear, for they were unmindful of army discipline, and every brave, in a measure, was sufficient unto himself.

Stirred as Peter was by the sight of the advancing army, he did not know anything of its leader, Rigaud, the brother of the governor of Canada, who was in command of this expedition of more than sixteen hundred French and Indians moving from Ticonderoga (or Carillon) to strike the little garrison of Fort William Henry. Regulars, Canadians, and red men were in the ranks, and they had been going leisurely on their way, strong, confident, and certain that there was no power to turn them back or prevent the capture of the fort. Even ordinary precautions were disregarded, and well supplied with weapons of offence, and protected from the cold and storms, they were apparently without fear. Overcoats, mittens, tarpaulins, moccasins, blankets, bearskins, axes, kettles, flint and steel, even needles and awls — these things and many more had been provided and strapped upon light Indian sledges, along with provisions that would abundantly supply all their wants for at least twelve days. They were not only no burden to the troops confidently advancing over the ice of the lake, but were an additional source of strength as well. A million francs had been spent in equipping the expedition, and even the men who had recently come the entire length of Lake Champlain over the ice were not wearied, for a week had been granted them for rest at Ticonderoga, and all were fresh and ready for the task before them.

Neither did Peter Van de Bogert know how poorly prepared the men at Fort William Henry were to receive or drive back the oncoming army. Not well equipped in any event, their unpreparedness had been increased by a carouse that had been held only the night before in "honor" of St. Patrick. The fort itself was not strong, and the weakness of its defenders was intensified by the night of revelry. The darkness into which Peter was peering was but a type of that resting upon Fort William Henry and its defenders.

So impressed had Peter been by the sight of the moving men, and so completely engrossed in watch-

ing them, that, for the time, even his own position
and the peril of his friends had been forgotten.
With a start he suddenly realized how recreant he
had been, and instantly departing from the shore
he hastened back to the hut. Running swiftly, he
entered the opening, prepared to give warning of
the advance of the army of Frenchmen; but, as he
gazed into the room, not a man could be seen.

Startled at the unexpected emptiness of the
hut, he was speechless for a moment; then striving
to convince himself that the darkness concealed
some who must be there, he called in a low voice.

Not a word was heard in response, and then,
frantically, Peter began to move about the room
as if with his groping hands he could grasp the
form of some sleeper who had been left behind in
the hut. But he could find nothing but a pair of
snow-shoes — his own he thought they might be —
and almost mechanically taking them, he once
more stepped forth into the night.

The silence was broken only by the whistling of
the March wind in the tree-tops, and not a man
could be seen. He strove to discover in which
direction the band had gone when they had departed, but it was too dark to enable him to perceive even their tracks in the snow. The loneliness
swept over him almost with a crushing weight.
The remembrance of the great body of men whom
he had seen moving over the ice only increased the

feeling. The very trees, leafless and high, seemed to him to sigh as if they shared in his dismay. For a moment he was tempted to cry aloud, to shout to his recent companions who apparently had fled without a thought for him. And yet as Peter realized how he had gone out from the hut with no word to any one, he was compelled to acknowledge to himself that he was the one to be blamed.

Resolutely he reëntered the hut, and seating himself on the ground strove to face his problem. Fort William Henry was not far distant, and his best move would be to attempt to make his way to it. His own help was needed by the garrison almost as much as he needed the aid of their presence. But could he make an entrance now? The body of Frenchmen had already passed on, and doubtless they had carefully made their plans to surround the fort. Even now other bodies might have taken positions assigned them and Fort William Henry be surrounded by its foes. Would he be able to penetrate the lines? The question was at least of doubtful solution. And yet to remain outside in safety, when such dire peril threatened his friends, was not to be thought of. Already, without doubt, Sam and his recent companions were well on their way, and were wondering what had become of him.

A brief time was sufficient to enable him to

decide. There was but one thing to be done, and
that was to strive to make his way into the fort.
The darkness was in his favor, and though the
attempt would be perilous, it must be made. He
had no weapon, his belongings had been lost, and
the only things that remained to him were the
two snow-shoes which he was holding in his hands.
With the thought that even these were not to be
despised and that he might find use for them, he
arose to his feet and advanced to the exit. But
as he peered out, the sight which greeted his eyes
was one that caused him quickly to withdraw into
the room and throw himself upon the ground close
to the opening.

CHAPTER VII

Pursuit

COMING from the midst of the near-by trees Peter could see a file of Indians. Even in the dim light he had no difficulty in perceiving what they were, and his instantaneous conclusion was that this must be a detachment from the main body of troops that he had seen moving over the ice. That some of them should have been assigned to the shore to discover any signs of danger that were to be found there seemed to him but natural, and the sight of the approaching red men confirmed his fears.

One glance had been sufficient to convince him that all attempts to escape would be worse than useless, for he would not be able to leave the hut without being seen. It might be that his presence had already been detected, and in the great fear that seized upon him he had instinctively withdrawn and in desperation stretched himself upon the ground near the entrance to the hut. The snow-shoes were still tightly clutched in his hands, although Peter could not have explained even to himself why he had retained them.

In a brief time he was aware that the red men

were drawing near the hut. He could hear their stealthy tread, and even the brief whispered conversation as they stood before the entrance was not lost, he was so near to them. Apparently the Indians were puzzled by the discovery of the recent construction, which was not unlike their own wigwams; but there were no signs of life within or about it, and if he had not been seen when he had approached the exit, there was a bare possibility that they might pass on without searching the hut or discovering his presence within it.

This brief hope, however, was speedily banished from his mind when he perceived that three of the men were stealthily entering. He could see the dim outlines of their forms as they pressed closely together and stepped inside. For a moment they stood motionless, peering intently into the room; and Peter, hardly daring to breathe, was in a fever of fear lest their eyes should turn upon his hiding-place, which was almost at their feet. One of the Indians, crouching low, advanced into the room, and Peter's fears almost overpowered him. What his fate would be if he was discovered he understood only too well, and if a search was made, he knew he must be found. He was pressing close against the brush that formed the outside of the hut, and his heart was beating so loudly that it seemed to him it must betray his presence. He could feel rather than see that the Indian was making his way to the opposite side of the room. What his purpose was he could not understand, but in a moment the man returned to his companions and with a low exclamation spoke a few words, and then all three withdrew.

Peter's heart gave a great throb of relief. Perhaps the man was convinced that the hut was empty, and all would now resume their journey, and departing from the place push on to join their fellows, who doubtless were already surrounding the fort. Even in the face of his own peril Peter felt a great throb of pity for the unsuspecting inmates of Fort William Henry, and he was wondering if his own recent comrades would be able to enter the fort or carry word to the garrison of the danger that threatened.

His thoughts, however, were speedily recalled to his own immediate danger, which was suddenly increased by the fact that the Indians had not departed, but were returning and were evidently all together just outside the hut. He was tempted to cry aloud when he perceived that a further search was to be made and that this time it was to be aided by a torch. He could see the shadows and glare of the fire already, and soon he could see the torch itself in the hands of one who was now standing just before the entrance.

Peter waited a moment until the man entered, and, holding the torch aloft, stood peering into the

room and gazing at the opposite side. There was
no hope of further concealment now, and acting
upon the sudden, wild impulse that seized upon
him, he grasped his snow-shoes tightly in his hands,
rose to his feet, struck the blazing torch, and with
a yell darted through the exit.

The unexpected blow had driven the torch from
the hands of the man who was holding it, and a
wild cry, almost like an echo of Peter's, broke from
his lips as he darted back among his companions.
Peter's one forlorn hope had been that in the
momentary excitement he might dart past the
little group; for he did not believe there were
more than four or five in the company, and if once
he could gain the shelter of the near-by woods, he
might have a meagre chance of escaping.

The startled red men had dodged back from the
entrance as Peter leaped among them, and putting
forth all his strength he bounded toward the forest.
Never before had he exerted himself as at that
moment. He dimly realized that there was a
momentary confusion behind him; but the protecting forest, although only a few yards away,
seemed to him to be too far to be gained. He
leaped with great bounds as he strove toward it.
The one loud shout he had given as he plunged
into the group, which had started back at the
sound and the unexpected appearance, was the
only demonstration as yet.

Just as he entered the forest he heard a gun
fired close behind him, but the excitement had
been too keen to permit the warrior to take careful
aim, and the bullet passed harmlessly over his
head. A wild shout of rage instantly followed, and
Peter did not require any glance to inform him that
all the men were now in swift pursuit of him.

Not once did he look behind him, for safety was
to be found only in the gloom of the forest. Onward he bounded, making great strides, and for
some reason still clinging to the snow-shoes which
he had held in his hands. He could hear the sound
of his pursuers, and knowing as he did the speed at
which they could run, he strove desperately to increase
his own pace. The snow was deep in places, and
in places light almost as feathers. He stumbled
and slipped and slid, and yet he never once stopped.
The sound of his own footfalls must reveal his
course to his pursuers, he thought, but he could
not help that. He vaguely wondered why they did
not fire at him again, as he was certain his form
could be seen; but for some reason they did not
shoot, and Peter Van de Bogert still ran on and on,
seeking no place, only striving to put as great a
distance as possible between him and his dogged
followers. He had been only a few yards in advance when he started, and his sole reliance was
upon the darkness and his own speed. The wind
cut sharply into his face, his cheeks were torn by

the low branches of the trees, his breath came in gasps; but still wildly clutching his snow-shoes he sped on, a faint hope beginning to dawn as he realized that as yet he had not been overtaken.

He could not, however, once glance behind him, for the ground over which he was running was rough and the snow was treacherous. To lose his foothold would be to lose his life. His sole hope of safety lay in pushing forward. Desperately, feeling every moment as if he was about to be seized from behind, his breathing labored, with a tightening of the cords of his throat, his eyes fixed upon the forest before him, he still held to his course.

The mad race, however, could not long be maintained at such a pace. Knowing as he did that for a contest of this kind the Indians were much better fitted than he, both by nature and training, Peter realized at last that if he was to escape it must be by some other means than that of mere swiftness. Thus far he had been able to hold his own, perhaps even to gain slightly; but his pursuers were close behind, and the moment his speed relaxed they would be upon him. Even the darkness did not provide the protection he had hoped for. Either it was not so dense as he had thought, or his eyes were becoming accustomed to it, and he well knew that the Indians could see much better than he.

Wildly he glanced about him for some means of escape. If he could only hide or discover some way of eluding his pursuers, his chance would be much better than by mere flight, he thought, but only the trackless forest spread out before him, and he could see nothing that would aid him.

Suddenly, through the trees he obtained a glimpse of the lake. Unknown to himself he had been following the shore, and was now approaching a little cove across which he must move. Once on the ice, he would be an open mark for the bullets of his enemies. Wildly he glanced on either side for some opening or place of refuge, but nothing could be seen; and then suddenly he slipped and fell, and before he could check himself, had rolled down the side of a small knoll and brought up sharply against the huge trunk of a fallen tree.

Almost stunned by the fall, for a moment he lay motionless, expecting his enemies to be upon him and that the end would then come. Already he could hear them, for they were regardless of the branches that snapped beneath their feet, so intent were they upon overtaking the fugitive. They could not be following the trail of his footprints in the snow; it was too dark to permit them to do that, Peter felt assured. Their keen ears had detected the sound of his movements, and had shown them the path to follow. Doubtless, after their custom, they had spread out in a long line; and knowing

the direction in which he was running, were certain
to take him if he should turn back, and to overtake him in case he kept on.   Either horn of the
dilemma to Peter seemed to be as bad as the
other, and yet for the present there was nothing
for him to do except to remain where he was.

When a brief time had elapsed and he had not
as yet been discovered, he suddenly bethought him
of his snow-shoes, which for some unaccountable
reason he had not cast away in his flight.   Without rising from the ground he reached out and
carefully and securely bound the shoes to his feet.
At every moment he expected to hear the report
of a gun or feel the grasp of an Indian, but as yet,
apparently, he had not been seen.   The darkness,
doubtless, had aided him, and the sound of his fall
had not been heard.   It might be possible, also,
he thought, that his pursuers had all passed the
spot where he was lying; but it would not be long
before they would return.   When he failed to
appear on the ice in the cove below, they naturally
would conclude that he had turned back on his
way, and would at once retrace their own course;
and Peter knew Indian nature too well to permit
him to hope that the chase would be abandoned.
Whatever he was to do must be done quickly,
and silently rising to his feet, he began to move
stealthily through the forest.

He had, however, advanced but a short distance

when his ears were saluted by a cry which he
knew was the signal of his discovery, and the
answering shout that came from behind him confirmed his fears. Close to his head passed a tomahawk, and before he could dodge he heard the thud
as it struck a tree beside him. Again the darkness had been his protection, but his enemies were
near, and doubtless a line had been thrown out to
intercept him if he should turn back.

The snow-shoes were at times of great assistance,
and then again he would pass over places where
they were a hindrance; but he could not dispense
with them now if he would. The wild flight had
been resumed, and his only hope lay in pushing
forward. At all events, he had not been taken,
and thus far he had escaped the weapons of his
pursuers, and with the thought came renewed
courage.

For a half-hour he kept swiftly on his way,
never once halting for rest or to look about him.
A new light had now appeared in the sky, and
Peter knew that the morning was at hand. Just
where he was he could not determine; but, suddenly, directly before him, he beheld a steep descent
in the hill, so steep as to be almost a precipice.

While he did not decrease his speed for a moment,
a fresh plan of escape flashed into his mind. As
he drew near the border of the declivity, he tore
his jacket from his back, and rolling it up sent it

down the hillside.   He could see it far below him
as it rested against a bank of snow.   Then, unstrap-
ping his snow-shoes, he reversed them as he rebound
them on his feet.   Stealthily, and with the utmost
care, he began to retrace his way, striving to plant
each footprint in the impress already made.   He
was in a fever of excitement, and every hope of
success depended upon the nearness of his foes.
He looked eagerly about him as he proceeded, and
in a brief time he came under the low branches
of a spreading oak that grew near a clump of
cedars.   It was but the work of a moment to
swing himself up, hastily remove his snow-shoes,
scramble from one branch to another, and then to
leap into the shelter of the thick cedars.

CHAPTER VIII

A Surprise

PETER VAN DE BOGERT settled himself in his place of refuge, rejoiced to secure even a brief breathing spell, for he was well-nigh exhausted by his efforts. The long journey which he had made with his companions, the effort to escape his friendly pursuers on the ice, the sleepless night, and later the wild flight from the Indians, had taxed his powers of endurance to the utmost.

In spite of his weariness, however, the plight in which he found himself did not permit him even for a moment to ignore the peril which still threatened. He knew that his pursuers were relentless and keen, and that the trick he had played might not avail. It was therefore with an eagerness still unrelaxed that he peered out through the branches of the cedars at the snow-covered ground beneath him. The light of the rising sun was becoming clearer, and the shadowy outlines of the great trees were now more distinctly seen. The very silence was made more appalling by the streaks of sunlight penetrating the depths of the forest. The air, too, was cold and biting, but Peter was almost

unmindful of that and of the loss of his jacket, so
eager was he in watching for the coming of his
enemies.

Nor had he long to wait. Out from among the
shadows came three Indians, following his trail
with all the intentness of a hound in pursuit of a
fleeing deer. His heart almost stood still as he
beheld them, and when they passed close to his
hiding-place, he could even discern the streaks of
paint on their swarthy faces. They turned neither
to the right nor left, and soon passed on, but they
had not disappeared before three others came
swiftly following them.

The brow of the declivity down which he had
cast his coat was not far away, and Peter could
easily see the men as they halted upon its border.
The three who were in advance waited for their
companions to join them, and then there was a
consultation. Their words were spoken in too low
a tone to permit Peter to hear them, nor would he
have understood even if he had heard; but their
gestures as they pointed at the coat which could
be seen at the foot of the descent, and then at the
imprint of his snow-shoes in the snow, left no doubt
as to the meaning of their consultation.

They seemed to arrive at the conclusion which
Peter had most desired, and for which he had
planned. They separated, three going swiftly to
the right and two to the left, and although they

were speedily out of sight, Peter naturally concluded that they were seeking the bottom of the
hill down which they believed he had cast himself
in his desperation. The chief difficulty now lay
with the one who still remained on the brow, evidently left behind to guard against any possible
mistake, and to be on the lookout for the reappear-
ance of the man they were seeking.

The solitary Indian stood gazing down into the
valley below, apparently with no thought but that
the fugitive would be seen somewhere there. The
moment was propitious, and hastily readjusting
his snow-shoes Peter swung himself down from his
hiding-place, and was again standing on the snow-
covered ground. He could no longer see the red
man, but knew he was near enough to hear the
slightest sound.

Using his utmost caution, Peter crouched low
and moved stealthily back in the direction from
which he had come. Every sense was alert and
keen, and every moment he expected to hear the
shout indicating that his presence had been discovered. Slowly and cautiously he withdrew,
carefully avoiding fallen branches, and stepping
lightly as he advanced. At last, when he had proceeded several yards and was convinced that his
movements had not been discovered, he broke into
a run. He turned sharply from the way by which
he had come, the one purpose in his mind now

being to place the greatest possible distance between himself and the half-dozen red men, who
would not require much time to convince them
that their prey had not cast himself down the hillside, as they had believed.

Every moment was precious, and Peter sped on
with all the strength he could muster. The snow
was deeper and also lighter than where he had
been before, and his flight was therefore swifter.
He abandoned all thought of direction, and plunged
forward through the great forest. Even his feeling
of hunger was ignored, and an hour had elapsed
before he halted to rest.

Then, for the first time, he listened for sounds of
pursuit, but the silence of the vast wilderness was
undisturbed. As far as he could see, the trackless
forest was devoid of all signs of life. The tall trees
swayed under the March wind, the sunlight that in
places cast its glow upon the snow was beginning
to be of a hue foretelling the approach of warmer
days; but neither man nor beast was to be seen.
Peter was alone, but the oppressive fear in his heart
did not long permit him to rest; and as soon as, in
a measure, he had recovered from his fatigue, he
resumed his flight, pressing forward rapidly, and
yet not so eagerly as when first he had doubled on
his tracks. That the pursuit would be abandoned
did not seem to him to be expected, and his sole
hope lay in striving to distance the tireless red men.

When he halted again, it was to think more calmly of his problem. He had been regardless of all direction up to this time, but now the necessity of striving to make his way to Fort William Henry reasserted itself. But where was the fort to be found? He glanced up at the sun and quickly decided that he must change his course and bear more toward the south if he would succeed. He had no conception of the distance, for, as has been said, all ideas of time or place had been abandoned in the wild flight he had made. The hour of noon would soon be at hand, and Peter knew that, at least, there could be but little time for resting now. His feeling of hunger, which was beginning to assert itself, strongly provided an additional incentive for action.

Drawing his belt more tightly about him, he resumed his way and kept steadily on until the sun was high in the heavens. Not a sign of his pursuers had been seen, and the young soldier began to hope that he had succeeded in eluding them. His immediate needs began to press more strongly upon him, but although he looked eagerly all about him, he was not able to discover anything that would relieve his hunger. His sole hope rested upon his ability to gain the fort before nightfall; and with the conviction came the resolution to make the attempt, although his peril was increased by the foes who might be now before him as well as behind,

and whose vigilance he must elude before he could
secure the shelter he was eager to gain.

His fears increased when the thought arose in
his mind that the little garrison might not have
been able to hold out against the great force that
had advanced upon it. Aware as he was how
greatly the numbers of the Frenchmen and Indians
exceeded those of the defenders, the hope that the
struggle could be prolonged seemed slight indeed.
And yet both duty and necessity urged him forward,
and, convinced that there was but one plan for him
to follow, he doggedly held to his way until the
middle of the afternoon. He became somewhat
puzzled by the fact that he had not seen any familiar landmarks, and for the first time the possibility
that he had wandered from his course began to
dawn upon him. The thought certainly was not
reassuring, but there was no time even for dwelling
upon that, for he knew that he must keep on
and on.

Suddenly he was startled by the sound of a dog
barking, not far in advance of him. It was so unexpected that Peter, for a moment, was tempted to
turn back, thinking that unawares he might have
approached an Indian village, and in that case the
peril would be as great as the one from which he
had fled. He was unarmed and defenceless, and so
utterly wearied by his exertions that he could not
think of attempting a second escape.

The barking of the dog became clearer, and in a moment Peter saw the animal breaking through the bushes. It was strange, he thought, that if he was near an Indian village, only one dog should be heard, for usually scores of them were to be seen; but he glanced hastily about him for a stick or club with which he might defend himself. The barking had ceased now and given place to low growls as the dog drew near, the hair on the animal's back standing erect and its manner betraying its suspicion of the stranger. Peter had not been able to find a club; but as he suddenly noticed that the dog before him was not like any he had ever seen in the villages of the Indians, he gave up all thought of defence, and, extending his hand and speaking kindly to the animal, waited for it to approach. The unexpected change of manner at once produced a corresponding change in the dog, for, whining and fawning, he drew near until Peter patted him upon the head and spoke to him kindly. This treatment caused him to rise with his fore-feet resting upon Peter and manifest every sign of delight, but in a moment he darted back in the way from which he had come, and barking joyfully disappeared from sight.

Somewhat puzzled by such strange actions, Peter at once decided to follow the animal. He was so faint that further flight seemed to be out of the question, and the conviction that the dog must belong to white men was so strong that, in his desperation, he was determined to test the matter before he proceeded farther on his way.

Advancing cautiously, he was surprised to find himself in a brief time on the border of what seemed to be a little clearing. A small house of logs could be seen, and near it was another structure, also of logs, which plainly served the purpose of a barn. For a considerable distance the trees had been felled, but the snow was too deep to enable him to discover whether or not the land had ever been cultivated. The dog, meanwhile, had disappeared, and Peter was not able to discover his presence by sight or sound. From the rude stone chimney in the side of the house nearest him, a thin curl of smoke could be seen rising, and it was clear that some one was inside the building, but whether friend or foe there was no means of knowing.

Peter's feeling of surprise became keener when he perceived that the heavy shutters upon the little windows were closed and that to all appearances the inmates were upon the defensive. It was evident, then, that they were in as great fear of him as he was of them; and acting quickly upon the thought Peter advanced into the clearing holding out the palms of both hands as he did so, after the Indian method of expressing that the purpose of his visit was peaceful.

He did not advance far from the border, however, still being uncertain as to his reception.
Although he knew that he presented an excellent
mark for any one who might be inclined to fire
upon him, he nevertheless hoped that his own
defenceless condition would be his best protection.
For several minutes he stood there, but there was
no indication that even his presence was known by
those who were within the house. The smoke
from the chimney had ceased, and Peter was
almost tempted to believe that his eyes had deceived him and he had not seen any when he had
first advanced.

"Hello! Hello!" he called. "Any one in the
house?"

There was the faint sound of the barking of the
dog in response to his hail, but that was all, and
even that quickly ceased. It certainly was strange,
well-nigh unaccountable, he thought; yet he was
somehow convinced that there were people inside
the building and that his own presence was known.
Many of the houses on the border were provided
with loopholes, and the feeling was strong upon
him that curious eyes were even then watching
him and that the muzzle of a gun might be
pointed toward him, although he was not able to
perceive it.

He waited a brief time, and then repeated his
hail, but the silence was unbroken even by the

barking of the dog. He was becoming desperate by this time, and resolved to approach the door. Peril there might be, but danger lay even in reëntering the forest, for he was weak and faint, and slight prospect of help was to be found there.

Advancing slowly, and striving by every means in his power to show that he was defenceless and that his purpose was peaceful, Peter moved across the clearing and again stopped when he was a few yards distant from the door. He waited, but still his coming seemed to be ignored.

"Hello! Hello!" he called again. "Will you let me in?"

This time the dog barked more loudly than before, but the sound ceased abruptly, and no one spoke in response to his hail.

Once more he began to move toward the house, but he had advanced only a few steps when suddenly the door was opened and a girl, or rather young woman, appeared in the doorway.

"Is that you, Peter Van de Bogert?" she called sharply.

"Sarah? Sarah Curtis? What are you doing here?" exclaimed Peter, in astonishment at beholding his friend.

"Come in, come in, Peter," she replied.

He hastily entered, and the door was quickly closed and barred behind him.

CHAPTER IX

THREATENING DANGER

IN the dim light, as Peter gazed about him in the room, he could see a rude bed in one corner, upon which lay a woman whose flushed face and general appearance at once proclaimed that she was suffering from illness. Three children, the eldest not more than eight years of age, were also in the room. Sarah herself, resolute and keenly excited, was the only occupant who seemed to be in any way capable of action; and as the young soldier looked questioningly at her, in response to his implied query, she said : —

"Yes, I'm the only one here besides my aunt and her children."

"I don't understand. I don't see how you — "

"I left home the same day you did," said Sarah, quietly.

"Why did you come? How did you get here?"

"My uncle came for me. My aunt was sick, and he wanted me to help, so I came."

"They ought not to have stayed here. Doesn't your uncle know — " Peter stopped abruptly, for he had no desire to increase the feeling of uneasiness

which doubtless Sarah already had.   The lonely
and somewhat exposed location of the cabin, the
defenceless condition of its inmates, were of themselves sufficiently alarming without adding to her
fears by the knowledge of the possible presence of
the savages.   He was puzzled, too, to account for the
absence of her uncle, who had no right to leave
the family at such a time as the present, he half
angrily told himself.   "You ought not any of you
to be here," he said aloud.

"What ought to be and what is are two different things.   We *are* here and must do the best we
can."

"Where is your uncle?"

"I don't know.   He went away this morning
and said he'd be back soon, but he hasn't come
yet.   You don't think anything could have happened to him, Peter, do you?" she inquired
anxiously.

"He probably knows what he is doing," replied
Peter, evasively; "but he ought not to have left you
here alone."

"I'm not alone.   You're here now, Peter."

"Yes," responded Peter, not insensible to the
confidence she displayed, and yet not entirely able
to disguise the feeling that he heartily wished they
all were somewhere else.

"You haven't told me how you happened to be
here, Peter," suggested Sarah.

"No, I haven't; that's so," he replied thoughtfully. He did not intend to tell her of his own escape or the dangers which, he was assured, had not entirely gone. "We spread out and agreed to go to the fort from different directions," he said at last. "We thought we'd be better able to find out in that way if any of the Frenchmen are in these parts."

"Did you find out?" she inquired quickly.

"Why, I don't think the region's entirely bare of them. It wouldn't be natural to expect that. How far is Fort William Henry from here?"

"Eight miles."

"Eight miles!" exclaimed Peter, aghast. "You don't mean that? It can't be as far as that."

"That's what my uncle said, and probably he knows."

"Yes, that's so; he probably does," said Peter, thoughtfully. "I must have gone a long way from the right direction."

"You look tired out, Peter. I ought to have noticed that before. And you must be hungry, too. Just wait a minute and I'll start up the fire and get you something to eat."

"No, no! Don't start any fire. I am hungry, but I'll eat anything you have. Don't try to cook anything now."

"Why not?" She was looking keenly at him as she spoke, and Peter knew that she had divined the source of his fear. However, he was still decided not to alarm her unduly, and so he said quietly: —

"It wouldn't be wise to do anything to call attention to us. I don't know that we have anything to fear, but it's better to be on the safe side, anyway; so if you'll just give me what you have ready to eat, that 'll be best."

"We haven't anything cooked but some cold corn bread."

"That's just what I want."

"How fortunate," replied Sarah, laughing for the first time since Peter's arrival.

She placed the bread upon the low table, and the half-famished youth fell upon it in a manner that would have rejoiced the heart of any cook.

When at last he arose from his seat, Sarah said: "Now, Peter, tell me all. You have more to say than you have told me."

"What makes you think that?"

"It doesn't make any difference. I want to know all about it."

For an instant Peter was tempted to tell her everything, but restraining the impulse as he quickly realized how useless it was to increase her fears, especially when her present anxiety was all that she ought to bear, he said: "There isn't any use in trying to hide the fact, Sarah, that no one knows what may happen. But it will be time

enough to 'know it all' when we have to. Just
now no one knows that, but to show you how I
feel myself, I'm going up in the loft and go to
sleep. I want you to keep watch while I'm up
there, and call me instantly if you see or hear anything. I don't want to be up there long, but I'm
so tired and am so sure I'll be in better shape if I
get a little sleep, that I'm going to be selfish
enough to do it."

"Shall I open the door?"

"No," said Peter, hastily; and then for fear that
he had unduly alarmed her, and also aware that
she could doubtless close and bar the door before
any one could enter, he added: "I don't know,
though, that it will do any harm to leave it open.
But if you don't have any fire, it will make it
cold for your aunt and the children. Do as you
wish."

He approached the bedside of the sick woman,
but she seemed to be sleeping, for her eyes were
closed, and she did not open them as he drew near.
Then, turning quickly, he mounted the rough ladder
that led to the bare room above, and disappeared
from sight. His heart smote him for what almost
seemed to him a cowardly act, yet he was nearly
worn out by his recent exertions. He knew, too,
that as it was only mid-afternoon the present peril
was not so great as that which might come a little
later, and he must fit himself, as best he could, to

face any danger. Besides, he had complete confidence in Sarah's watchfulness, and there was also the possibility that her uncle might return at any moment. If he did not, Peter had already decided to remain in the cabin for the night, for he would not leave its inmates without protection of any kind. It was true that he was long past due at Fort William Henry, but the recollection of the body of troops he had seen moving over the ice of Lake George was still fresh in his mind, and he believed that by this time the fate of the little garrison was already settled. Aid might have come from Fort Edward, or the garrison might have been cut off before such aid could arrive; but in either event his duty for the present seemed to be clear, and with this conviction in his mind he was soon asleep.

He had no conception of the time which had elapsed when he was aroused by the touch of Sarah's hand upon his arm. Quickly sitting erect, he rubbed his eyes in confusion, and then the recollection of the recent events and the place in which he was flashed upon him.

"What is it, Sarah? Have you seen anything?" he demanded, as he instantly rose.

"I don't know. I can't tell. Come down and see."

Thus bidden, Peter hastily descended the ladder, and Sarah followed him. As soon as they had

gained the floor below she led the way to one of
the small loopholes with which the cabin had been
equipped, for every settler fully understood that
his house must be his castle in more ways than
one. Peter peered out as his friend had bidden him.

For a time he was not able to perceive what it
was that had alarmed her. The leafless forest beyond the border of the clearing stood out, sharply
defined, against the ice and snow. The low barn
of logs, the charred stumps that still remained in
the clearing, the long well-sweep, were the only
objects he could see, and no human being, white
man or red, apparently was within sight. He
could see for a considerable distance down the
rough road or lane which had been cut in the
forest, and by which the occasional traveller
approached the lonely abode; but that, too, seemed
to be devoid of all living objects.

"What was it, Sarah?" he turned and whispered
to his companion, who was standing anxiously by
his side. "I don't see anything. Where was it?"

"Look straight beyond the corner of the barn,
right there where the lane begins. Watch those
big stumps on each side, and then tell me if you see
anything."

Thus bidden, Peter did as he was directed, but
still he was unable to make out anything, until
from the depths of the largest hollow stump he
perceived a head rise slowly, and instantly recog-

nized it as that of an Indian. The head disappeared from sight, nor did Peter see it again, although he watched carefully for several minutes.

Then, partly turning, he whispered to Sarah, "How many guns have you here in the house?"

"Two."

"Muskets or rifles?"

"One rifle and one musket."

"Bring them both here."

The girl did not leave her place, but bade the oldest of the children bring both guns and also the powder-horn and bullet-pouch. As soon as these were at hand, Peter said: —

"Now, Sarah, I want you to take my place and watch that stump."

The resolute girl did as she was bidden, while Peter loaded both guns with all haste and set them where they would be within reach as he once more took his place at the loophole. A groan from the sick woman sent Sarah at once to the bedside; and a wail from one of the children, who all seemed to be aware now of some peril threatening them, caused her, as soon as she had given her aunt the water she had called for, to stop and say to the oldest: "Now, Lucy, you must not let the children cry! Take the bean bags and the rag dolls and play with them; but don't you cry, whatever happens! I know you will do this, for you can help a good deal."

The little girl, after the serious manner of the pioneer children who were early made " to bear the yoke," at once complied with the request, and Sarah returned to Peter's side.

"Why don't they attack the house, Peter?" she whispered.

"Can't tell yet.   It's strange, but we'll know pretty soon.   Did you see more than one?"

" Yes, I saw three."

" Where ? "

"One was inside the hollow stump, and the others were on the other side right behind those cedar bushes on the left."

"I haven't seen them.   They're bold to show themselves."

" You don't suppose they're waiting there for my uncle to come, do you? "

Peter had already thought of that as a possible solution of the positions the Indians had taken, and their apparent fearlessness of danger from the direction of the cabin also argued that they were not altogether ignorant of the inmates and their defenceless condition.   If that were true, then, he reasoned, these men in all probability were not the ones who had pursued him, for he was convinced that they must have gone on with the army of Frenchmen.   But whoever these men were, their purpose in hiding as they did could not be a friendly one, and he must be prepared to act the moment their hostile intentions became apparent.

He glanced at the sun and could tell that within an hour night would be at hand. He looked again at the stump and the bushes, but not a sign of life was to be seen.

"Sarah," he suddenly whispered sharply, "please take my place for a moment."

As soon as she had followed his directions he crossed to the opposite side of the room, blaming himself for his negligence, and hastily searched for signs of the enemy from that quarter. He carefully scanned the borders of the forest and looked keenly at the blackened and disfiguring stumps, but he had seen nothing to alarm him, when he was sharply recalled by a low exclamation from Sarah. Recrossing the room he took her place, aware from her manner that something was wrong, and his first glance confirmed his fears.

The Indian who had concealed himself in the hollow stump had raised his head and shoulders above the top, and thrust out a gun which was pointed directly down the roadway. One quick glance in that direction enabled Peter to perceive that some one was approaching, and that doubtless the Indian's delay was due to his desire for the man to come nearer before he himself acted.

CHAPTER X

WATCHING

QUICKLY, and yet carefully, Peter thrust the barrel of the rifle through the loophole and aimed at the head which he could see just above the edge of the stump. His heart was beating furiously, and he could almost feel the excitement under which Sarah also was laboring. There were no misgivings or compunctions in his mind, for the tales of the massacre at Schenectady were familiar in all the neighboring settlements, and the well-nigh universal feeling against the red men was that of hatred. They were treacherous, cruel, bloodthirsty, and although doubtless some amongst them were not so bad as others, the negative consideration was the only one bestowed upon them. Even the Mohawks, or the Five Nations, were not to be relied upon, and the death of King Hendrich, nearly two years before this time, in the battle of Lake George, had lessened the zeal of his warriors, at no time keen for the English settlements within recent years.

Peter Van de Bogert's feelings, therefore, were not unlike those by which all his neighbors were animated, and of their justice he had no question. His present action, however, was governed by the sole desire to save the approaching white man from the peril which was threatening him and of which he could have no knowledge.

For a moment the head and shoulders of the red man rose a little higher from the stump, and Peter knew that the time for action had arrived. He took careful aim and pulled the trigger. The report seemed to be doubly loud, but for a moment the puff of smoke hid the sight of his target from his eyes. His ears, however, were saluted by the sound of the discharge of the Indian's gun, and this was instantly followed by such a yell as it seemed to him he never before had heard. More than the three savages whom Sarah had seen were evidently in hiding, and the startling shot had for the first time roused them from their cover. The cloud of smoke passed, and Peter could see several of the red men as they moved about in the border of the forest. He seized the musket and again fired at the nearest man, and once more the wild cry rose from the border of the clearing; but instantly every Indian disappeared from sight, and a silence deep and intense rested over all.

Peter had no means of knowing whether he had hit the warrior or not, or if the Indian's shot, which had been fired almost at the very instant when his own finger had pressed the trigger, had found its

mark. The sun was just disappearing from sight
in the western sky, and the night would soon be at
hand.

He reloaded the guns, Sarah meanwhile keeping
watch, and then he resumed his place at the loophole, bidding her, at the same time, to take her
stand on the opposite side of the room and report
if she discovered any signs of peril from that direction. He was sufficiently familiar with the ways
of the Indians to know that they were not likely
now to depart from the place, and that if the Indian at whom he had first fired had been hit, his
comrades would at least make an effort to take
him, living or dead, with them when they left the
clearing. He could see the stump distinctly, but
had not discovered a sign of the Indian since the
shot. He might have been hit and fallen back into
the hollow, but of that Peter could not be positive.

A half-hour elapsed, and the silence was still unbroken. The dusk was beginning to deepen, and
yet the vigil in the little cabin was maintained.
The moaning of the sick woman had ceased, and
even the children were silent, so strong was their
feeling of terror. Sarah had withdrawn from her
place at the loophole long enough to obtain and
divide some of the corn bread among them, and
then had hastily resumed her watch.

Suddenly, Peter was positive that he could see
the figure of a man emerging from the forest in the

direction of the barn. He crouched low and ran swiftly from the border to the shelter of a huge blackened stump, not more than five yards away from the place he had abandoned. Peter's first impulse had been to fire; but he had hesitated, for in the dim light he could not be positive that it was an Indian, and the fear of a mistake held him back. He kept his eyes fixed upon the stump, but several minutes elapsed before the man moved from his hiding-place. His actions, as he crept stealthily toward the nearest stump, still moving in the direction of the barn, caused Peter to call to Sarah in a low voice, and say, when she came in answer to his summons: —

"Sarah, there's a man hiding behind that big stump right in line with the corner of the barn. I've watched him, and I think it's a white man. You try now and see if you can tell when he dodges out again, as he will in a few minutes. It may be your uncle."

Sarah obediently took her stand at the loophole, and in a brief time she whispered excitedly: " I've seen him. He's crept up to that stump that is higher than any of the others. It's right out there," she added, pointing as she spoke to the tall stump of a huge pine tree a few yards nearer the barn than the one where Peter had last seen the man.

" Is it your uncle ? " he whispered.

"I think so, but I can't be sure. It's a white man, I should say, from his actions."

"He's trying to get to the barn," suggested Peter.

"Yes, that's plain; and I think it must be my uncle."

"Has he got a gun?"

"He had something in his hands that looked like one. I've no doubt he has, for he took one when he went away this morning."

"There! There he goes again!" whispered Peter, eagerly, as the man could be seen darting toward another stump. "If it is your uncle, I should think he'd know enough to wait till it's dark before he tries it."

"It's almost dark now," suggested Sarah.

"Not dark enough, though. He's gone about half the way. Maybe he'll make it after all. I wish — "

Suddenly Peter stopped, for the man had once more appeared and had started for the shelter of another stump; but this time a yell arose from the border of the forest, and in a moment several shadowy forms could be seen darting from amongst the trees and running swiftly after him. The fugitive was no longer crouching, but was running at his utmost speed toward the barn. All caution had been abandoned, and he was intent only upon gaining the shelter of the building.

"It's my uncle!" exclaimed Sarah, excitedly. "Shoot, Peter, shoot!"

Peter was as intensely excited as she, but he made no response.   He was watching the race, and waiting for the men in pursuit to come nearer. The white man had the advantage of a good start, but his pursuers were much fleeter than he and were gaining rapidly upon him.   On and on sped the racers, when suddenly the man in advance slipped as his feet touched a patch of ice, and he fell sprawling to the ground.

In an instant he had risen and resumed his flight, but an exultant yell from his pursuers had greeted his mishap, and a groan escaped the lips of the watching and terrified girl.

"Shoot, Peter!" she begged.   "Why don't you shoot?   Here, let me take the gun, and I'll do it if you are afraid!" she added, almost beside herself with terror.

Apparently ignoring her pleadings, Peter still did not fire.   He was watching the chase with breathless interest.   It was plain that the white man had not been disabled by his fall, but it was equally clear that his pursuers were gaining swiftly. That he would reach the shelter of the barn before they could overtake him was probable; but just what the man hoped for, even if he entered and secured the heavy door behind him, Peter could not understand.   He would be nearer the house, it was true, but there were ways of driving him from his retreat which Peter clearly understood the savages might adopt, and his heart sank as he realized what these would mean for the unfortunate man. No, the only hope rested upon the coolness with which he himself should act, and resolutely he strove to repress his excitement, which was as keen as that of the girl by his side.

"Peter! Please, Peter!" begged Sarah, her voice breaking as she spoke. "Let me use the rifle if you won't. I can't stand it to see my uncle —"

Her pleadings were interrupted by the report of the gun in Peter's hands, and then, almost savagely, he seized the musket which Sarah was holding and discharged that also.

A wild yell rose at the sound of the guns, but it was too dark for Peter to perceive just what the effect of his shots had been. In a brief time he saw that the white man had gained the desired shelter, but his pursuers were nowhere to be seen. The door of the barn faced the little cabin, and in the daytime was distinctly visible from the house. In his flight the man had turned the corner of the barn, but instead of keeping on toward the house had darted into the building and closed the door behind him.

Unable to understand why he had not kept on to the cabin, and fearful that he might still think his pursuers were close upon him, Peter shouted: "Come on! Come on into the house! You can make it now! Come on! Come on!"

But no response came from the barn, nor was the door opened.

"I'm afraid he's hurt," suggested Sarah. "I'm going out there."

"No, you're not! You're going to stay right here where you are," said Peter, sharply. "I want you to keep watch while I'm loading up the guns again."

Peter reloaded the guns, Sarah obediently taking her place at the loophole until her companion insisted upon resuming his watch, sending her back to the other side of the room, where she had before been stationed. The room was cold now, for no fire had been kindled in the fireplace; and stepping back a short distance from the opening, Peter said:

"Don't leave your place, Sarah, but tell me if there is wood enough in the house for a fire."

"Yes, but what do you want of a fire?"

"To keep you and your aunt and the children warm, for one thing."

"What else? You're not afraid to have a fire, Peter?" she asked.

"Not now, and it 'll be a protection, too, in other ways."

At Sarah's word the children dragged wood to the fireplace, and then under Peter's direction the intrepid girl abandoned her post for a brief time and soon had a blazing fire on the rough stone hearth.

"That's better," said Peter, "only you don't want to stand directly in front of your loophole now. Those redskins are not going to stay out there among those stumps all night, and if they get near the house, you don't want to furnish a mark for their bullets."

"Peter, are you afraid?" said Sarah.

"'Afraid'? Of course I'm afraid, but I'm going to watch all the more just because I am afraid. Now you mustn't talk any more, — that is, unless you see something, and then you must let me know right away. I'm watching the door of the barn, and am on the lookout for your uncle. He may start for the cabin any minute, and I must be ready."

"If it is my uncle," said Sarah, thoughtfully.

"I thought you said it was."

"I said I thought it was, but I wasn't sure. If it was, we'll know it soon."

Silence again rested over the little clearing and was not broken within the cabin, save by an occasional deep breath on the part of one of the sleeping children or a moan from the bed where the sick woman was lying. Peter's anxiety had increased as the darkness deepened, for he well knew that the designs of the Indians would be aided by the night. And in his heart he was positive that he knew just what these were. But forewarned was forearmed, he assured himself,

and resolutely he watched the barn and the part
of the clearing that lay beyond it, confident that
Sarah would guard her side well.

He had expected that the man who had gained
the shelter of the barn would make a break for the
house, but the slow hours dragged on and not a
sign of his presence had been seen. It was strange,
Peter told himself again and again, but the assurance afforded no solution.

The moon had risen now, and in its dim light
the outlines of the low barn and the disfiguring
stumps could be more plainly seen. The silence
of the night was made more impressive by the
moonlight, but the lonely vigil was still maintained
until it was long after midnight and the dawn
was not far away.

CHAPTER XI

Waiting for the Morning

IN the midst of the darkness, which seemed denser as the morning hour drew near, Peter suddenly perceived a tiny flame leap up from one corner of the barn. It was the sight which he had been expecting, but when it appeared his excitement and alarm were as keen as if he had had no thought of its occurrence.

"Sarah!" he called in a low voice.

The girl at once abandoned her post on the opposite side of the room and came to the place where Peter was standing. No explanation was needed, and for a brief moment she was speechless as she gazed at the appalling sight.

"Is there anything we can do, Peter?" she inquired anxiously.

"Nothing, except to watch."

"But that won't do any good!" she said impulsively.

"It's all. If we tried to go out there and put out the fire, we'd only expose ourselves and make a good target for the fiends. I —"

Peter was interrupted by a yell that rose from a place just beyond the barn. No one could be seen, but the shout was its own interpreter. From the sound it was evident that five or more of the Indians were there, and that they were now confident that their patient waiting was to be rewarded.

"My uncle! My uncle is there in the barn! He'll be burned to death!" exclaimed Sarah. "We must do something. He'll be shot if he tries to run to the house, and will be burned up if he stays there. Oh, what shall we do? What shall we do?" moaned the terrified girl.

As if to make matters still worse, one of the children awoke and began to cry. At the same time the yelling of the savages was renewed, and the flames began to make strong headway in the burning barn. They leaped above the low roof, and all about the place it was almost as light as day.

"Look after the children, Sarah," said Peter, briefly.

The child was quieted at once, and Sarah resumed her place by Peter's side.

"Is there a trap-door in the roof?" he inquired.

"Of the barn? I don't think there is."

"No, no. Of the house."

"Yes. Are you afraid of —"

"I don't know. I must keep watch. You never can tell."

Peter had been anxiously watching the door of the barn, expecting every moment to see the man who had sought its shelter break forth and start for the house.   He was aware what such an attempt would be likely to mean, and was prepared for what would follow, but he had not mentioned his fears to Sarah.

"Why doesn't he come?   What makes him stay there?   He'll be burned alive!   Oh, why doesn't he come?" said Sarah, sobbing as she spoke.

"He may have left the barn in the night," suggested Peter, although he had but slight hope that his suggestion might be true.   It certainly was strange that the man had made no attempt, as far as he could see, to leave his shelter, and Peter had kept a careful outlook all night long with this very thought in his mind.

The flames now were roaring, and the corner of the barn was all ablaze.   Sparks, too, were borne by the wind far out from the low building, and some came in the direction of the house.   The meaning of the question as to the trap-door in the roof was apparent now, and in his increasing fear the young man said to the girl by his side : —

"Sarah, you stay here a minute.   If your uncle leaves the barn, and an Indian shows himself, use your rifle.   I'll be back in a second."

Leaving Sarah at the loophole, Peter hastily mounted the ladder and entered the loft.   The trap-door was in the rear, near the stone chimney, Sarah had explained; and making his way to the place, he speedily lifted the heavy door a few inches and gazed fearfully all about the roof of the building. With a sigh of relief he let the clumsy contrivance fall back into its place, and with all haste returned to the room below. Before he could regain his position at the loophole his ears were saluted by a yell louder than any that had before been given, and this was speedily followed by the sound of a shot.

He rushed to the open place and gazed out, hardly aware for the moment that the girl was not where he had left her. But the sight that greeted his eyes was one that almost caused his heart to stop beating. In the light of the flames he could see Sarah herself running swiftly toward the barn. The fleet-footed girl stopped before she entered, and Peter watched her bend low over what appeared to be the body of a man. Then he saw the figure rise and Sarah assisting him to stand. Slowly, or so it seemed to the excited watcher, they began to make their way toward the house.

There had been a silence after the shot had been fired, and Peter was now expecting to hear the sound repeated or to see some of the savages leave the shelter of the stumps and start in pursuit. He held his rifle in readiness and hardly dared to breathe as he watched the girl and her companion. They had advanced but a few steps when he beheld a form dart out from behind the barn and steal up behind the two. In his hand the warrior was hold-

ing a tomahawk, and as he bounded forward Peter
expected to see him throw the weapon at the brave-
hearted girl.

Instantly he fired, and as the Indian fell forward
upon the ground, Peter replaced his rifle with the
musket which was by his side.   A fierce yell had
greeted his shot, and as he saw the red man rise to
his feet and dart behind the nearest stump he realized that though the pursuit had been checked, all
danger was not yet gone.   His fears were confirmed, when from the stumps on the opposite side
of the barn another shot was fired, and he gazed
at Sarah and the man, expecting to see one of them
fall.

Both, however, were struggling forward, and
now were within a few yards of the house.   A few
more steps and they would gain its shelter.   In an
agony of suspense Peter cast aside his musket, useless now, and darted to the door.   Holding it partly
open, he shouted : —

"You're almost here now.   Just a little more.
That's it !   That's it !"

As if in response to his call the fugitives increased
their speed, and stumbling, falling, at last staggered
through the opening Peter made, and the door was
once more hastily closed and barred.   Before it
was in place, however, the yelling of the savages
was redoubled, and the thud of their bullets was
heard as they struck the logs close to the doorway.

Sarah and her companion were safe, although the man had not risen from the floor where he had fallen.

Before he assisted him, Peter's first task was to reload the two empty guns; and then he turned to help Sarah, who was bending over the prostrate form. The light was clearer now, for the dawn had appeared, and as Peter looked down into the face of the unconscious man, he said quickly:—

"Who is this man, Sarah?"

"I don't know."

"I do."

"Who is it?"

"His name is Timothy Buffum, and how he came to be here beats me."

There could be no doubt as to who it was, and though Peter was mystified by the unexpected sight of Buffum, when he had expected to see Sarah's uncle, he quickly returned to his place at the loophole, leaving his friend to attend to the wants of the sufferer.

At once his entire attention was absorbed by the burning barn. The low building was now a mass of flames that were roaring and leaping high in the early morning air. The sparks, too, were lifted by the wind and carried in all directions about the little clearing. The sight was fascinating, but Peter was not thinking of its impressiveness. His face was haggard, and the long strain through which he had been passing was beginning to tell severely upon him.

Still there was no one upon whom the inmates of the house could depend except himself, and resolutely he strove to meet the demands upon him. What the end was to be no one could foresee. That not more than five were in the attacking party was evident, both from the numbers he had seen as well as from the shouts they had raised. If there had been more in the band, the house as well as the barn would surely have suffered at their hands. They doubtless were in ignorance of the number within the house, and this fact, as well as the result of Peter's shots, had kept them at a respectful distance from the building.

A shower of sparks flew directly over the roof, and Peter turned to Sarah, saying: " I must go up to the trap-door again, and you must watch here while I'm gone. Is the man dead ? "

" No, I don't think so. He was hit, though, I'm afraid. I don't know how badly off he is yet."

" You'll have to leave him. Be quick ! "

Peter rushed up the ladder again and made his way to the trap-door, which he partly opened. The sight which greeted his eyes caused him instantly to drop the door into its place and dart to the room below. A huge spark had fallen on the roof, and already was beginning to burn its way into the timber.

His feet hardly seemed to touch the rounds of the ladder as he descended, and the moment he gained the floor he said breathlessly : —

"Is there any water in the house?"

"Just one bucketful."

"Is that all?" he groaned. "Get it! Get it, for your life! The roof's on fire!"

The terrified girl did as she was bidden, ignoring now the cries of the frightened children. In one corner of the room when Peter had first entered he had noticed a short fishing pole, and this he instantly secured. Sarah rushed back with the wooden bucket partly filled with water, and as she placed it at his feet, he fairly shouted, "Get me an old coat, a dress, anything!"

Darting to a wooden chest, Sarah lifted the lid and drew forth a gown which had been the one piece of finery in the possession of her aunt. There was no hesitation, however, for instantly the dress was bound by a leathern strap about the end of the pole and then thrust into the bucket of water.

"Go back to the loophole!" called Peter, as with the pole in his hand he began to clamber up the ladder. "Shoot the first redskin that shows his head!"

Only partly opening the trap-door, Peter thrust forth his dripping pole, and to his delight perceived that he was just able to reach the spot where the

fire was. Dark smoke was already beginning to
rise from the timber, and he knew that he had not
a moment to lose. There was a sharp, hissing
sound as he touched the fire. He rubbed the
wet garment hard upon the place, and the fact
that the roof itself was wet from the ice and snow
of the winter favored his efforts.

He rushed down to the floor, thrust the
end of the pole into the little water that was left,
and went back to his task. This time his efforts
were rewarded with success, and he was satisfied
that for the moment the threatening peril was
averted. But the barn was still burning, and his
heart was by no means light when he returned
to the room below at the thought that nearly
all the water had been used.

"Look to your aunt and the man now," he said,
as he took the rifle from Sarah's hand and resumed
his place at the loophole. "Have you seen anything?" he added.

"Nothing."

"That's good. I believe the fire in the barn
is getting lower, too," he added, as he glanced at
the burning building. As he spoke the roof of the
barn fell in and a shower of sparks rose high into
the air. "Here, Sarah," he called, "I must go up
to the roof again, and you must stand on guard
here."

Placing the rifle in her hands, he once more

hastened to the trap-door and looked out over the roof. He groaned as he perceived that in a half-dozen places burning embers had fallen, and that there was every prospect that the roof would soon be on fire. And there was no water in the house with which to fight it, while the well, with its long sweep, was at least twenty feet from the rear door. To draw the water from its depths would require more time than could be had, and the attempt itself would be extremely perilous for the one who should venture forth from the house.

Almost in despair he looked about him, and then quickly decided that but one plan remained. He must crawl out and with his own hands strive to put out the blazing embers. Not once did he glance below as he quickly clambered out upon the roof, and lying flat upon it, began to crawl toward the nearest of the threatening places.

CHAPTER XII

An Unexpected Discovery

THE light from the burning barn was still strong all about the clearing, and Peter Van de Bogert was fully aware of what would happen if the besiegers should perceive him on the roof. The danger to the inmates of the house, however, was even greater from the fire than from the savages, and resolutely he crawled toward the nearest of the burning embers. He crouched low and pressed steadily forward. So intense was his desire and so threatening was the immediate peril, that, almost unmindful for the time of the presence of his enemies, he was intent only upon throwing the embers from the roof.

Soon he grasped the nearest of the blazing brands and hurled it to the ground. Even the pain which the hot wood produced was forgotten or ignored, and he began to move swiftly toward the next of the places threatened. He had advanced but a yard when a yell informed him that his presence had been discovered. A bullet struck the roof near him, and in the momentary panic that seized upon

123

him he turned sharply about, determined to seek the shelter of the interior of the cabin.

The snow and ice upon the roof had been melted only in places, and in his eagerness he was unmindful of his position. Suddenly his hands and feet slipped as he strove to hold himself in one of the icy places, and in a moment he realized that he was sliding from the low roof. Desperately he endeavored to clutch the wood beneath him. He tried to drive his feet into it or to grasp it with his hands. But the momentum with which he was moving could not be checked, and in a moment he shot over the low projecting edge, feet foremost, and struck upon the ground.

The fall was not far, but his descent had been seen by the Indians, and again he was fired upon; but the savages, never good marksmen, were not able to hit a moving target, and once more he escaped. The moment that he struck the ground and bounded to his feet his mishap was greeted by another wild shout, and as Peter glanced for a moment in the direction from which the yell had come, he perceived two of the red men advancing from the shelter of the stumps and running toward him. For the moment it seemed to him that all was lost. He felt as sometimes he had when in his dreams he had been struggling to escape the clutches of a wild beast. His feet seemed to be held back by great weights; he was carrying a

"IN A MOMENT HE REALIZED THAT HE WAS SLIDING
FROM THE LOW ROOF."

heavy load beneath which he was being crushed; even his ability to breathe was being shut off, and he was unable to move hand or foot.

Yet he was running with incredible swiftness all the time. It was only a short distance to the door in the rear of the cabin, and this he was striving desperately to gain before his pursuers should be upon him. For a moment, as he turned the corner of the house, the sight of them was cut off, and just then a shot rang out from the place where Sarah was on guard. The intrepid girl, aware of his peril, had done all that lay within her power.

The report of the gun was answered by a shout which told Peter that the pursuit had not been abandoned. Did they know just how many were within the house? And that the shot was fired by a girl? Even in his breathless flight the questions flashed into Peter's mind. But his efforts were not relaxed, and in a brief time, which seemed to him like long hours, he arrived at the door he was seeking.

Throwing himself against it he strove to open it, but it was fast shut and barred.

"Sarah, Sarah!" he shouted. "Open the door! Open it! Open it!"

He could hear the movements within, but the delay in coming seemed to him to be almost endless. He again threw himself against the door and endeavored to push it in, but the woodwork was heavy, and he was able to make no impression upon it. The clearing was silent, but Peter's fear was every moment becoming greater. At any time now the hideous forms of his foes might appear around the corner of the cabin, and he was unarmed and powerless to defend himself.

He was about to leave his place and run to the opposite side of the building, when suddenly the door was opened part way, and instantly he darted within and closed and barred it behind him. In reality the movements of Sarah had been carried out with wonderful quickness, although to Peter the intervening time had seemed almost endless.

For an instant he listened intently to discover if his pursuers would make any attempt to batter down the door, but the silence was complete. Then he ran to the loophole and, seizing the musket, peered out into the clearing.

The light of morning was now upon them, and he could plainly see all before him. The barn was still burning, but the danger of flying sparks was mostly gone. The blackened stumps appeared even more unsightly than on the preceding day, but Peter Van de Bogert was fearful of the peril from the roof, for he had failed to cast all the burning embers to the ground, and as yet he had no means of knowing whether or not the ice and snow would serve as a full protection from the fire.

Sarah had quickly reloaded the rifle while her

friend had been watching with the musket in his hands. The weapons were now exchanged, and after a brief silence, Sarah said in a low voice : —

"Did you put out all the fires, Peter?"

"No."

"What are we to do?" Sarah's voice was still low, but her tones could not entirely conceal the terror that seemed to possess her.

"We must wait a few minutes," said Peter. "It may be that there's ice and snow enough on the roof to keep it from catching."

"If there isn't?"

"Then I'll have to try it again."

Neither spoke for a time, and both were listening intently for the sound they dreaded to hear. The woman on the bed was moaning again now, and Timothy Buffum lay stretched upon the floor on a blanket near the fireplace. He did not stir, and Peter had no means of knowing how badly he had been wounded, nor was there any opportunity at present to investigate.

"Sarah," he said, breaking in upon the tense stillness, "I think you'd better go up in the loft and watch. If the fire should get hold of the roof, you'd know it, and you must call me the minute you find it."

Without making any reply she climbed the ladder and disappeared into the upper room, while Peter continued his guard.

The slow moments dragged on for some time. Not a soul had been seen among the blackened stumps, nor had any one approached the house. Peter was convinced that his recent pursuers had kept on in their course until they had entered the forest; and then, without doubt, they had returned to the place where they had left their companions, keeping all the time within the shelter of the surrounding trees.

He must know, however, that all was well on the opposite side of the cabin, and darting across the room he looked out cautiously. The blackened stumps were here as they were on the other side of the clearing, and not a living object could be seen. He strained his eyes to see into the borders of the forest, but apparently the great trees were not the hiding-place of any enemy. He was about to turn back and resume his place when he was startled by the sight of two men approaching from the forest. His first thought, that they were a part of the attacking party, was speedily banished when he was convinced that both were white. And their bearing, and apparently careless manner of approach, indicated that they were unaware of any peril to be faced in the little clearing.

Peter instantly thrust the muzzle of his rifle through the loophole and, pointing it into the air, fired. The report had the effect he had desired, for the men halted, and for a moment glanced keenly about them as they sought the shelter of the great trees. Both had disappeared from sight; and convinced that he had warned them of their peril, he quickly reloaded the gun and prepared to resume his watch at the place where the greater peril was to be feared. He had not been able to determine who the men were, or whether they were Frenchmen or not, and in the presence of the double uncertainty his anxiety was greatly increased. The fact that the men were white, however, was something for which he was grateful, for even if they should prove to be Frenchmen, surely they could not refuse to assist a household stricken as was that which they were approaching.

"Peter, Peter!" suddenly called Sarah from the upper room.

"What is it?"

"I'm sure the roof is on fire in one place at least."

For a moment the young man did not know what to do. To desert his place might be to give an opportunity to the red men who, he was positive, had not relaxed their vigilance, and to attempt to fight the fire seemed well-nigh hopeless.

Calling to Sarah to take his place, he mounted the ladder, and soon saw that his friend had spoken truly, for smoke was to be seen in one place, and the odor of burning wood was distinctly perceptible. To venture upon the roof would be to expose

himself to a peril greater far in the clear light of
the morning than it had been in the dimmer light
when he had made the attempt before.   There was
no water in the house, and none to be obtained
without venturing to the well, and there the danger was even greater than upon the roof.

Something must be done, and at once, he decided,
and he slightly raised the trap-door.   He could see
the spot where the fire was beginning to make
headway, but it was near the edge of the roof, and
so far beyond his reach.   He looked about the
clearing, but although he was not able to see any
of the Indians, he did not doubt that their keen
eyes were watching for the appearance of some of
the inmates of the house.   He glanced, in despair,
in the direction where the two white men had
been discovered, but neither was now to be seen.

Lifting the trap-door a little higher, he turned
his face toward the side of the clearing where they
had appeared, and shouted : —

"Come and help us ! The house is on fire !
Come !"

A head partly appeared from behind one of the
largest stumps, and a voice called : —

"How many in the house ? "

"I'm the only man."

"Redskins hereabouts ? "

"Yes."

"How many ? "

"Five or six. I don't know just how many. Come! Be quick!" Peter could see that the smoke was becoming denser, and soon the blaze would break forth and it would be too late to save the building.

"Are they on this side or the other side o' th' house?"

"Mostly on the other side. Some may be here, but there can't be many. Are you coming? Come on! Be quick about it! We haven't a minute to lose!"

Peter was in an agony of apprehension now, for with white men near, and speaking English as they did, they were doubtless friendly. And their friendliness, to be of service, must be shown at once. Besides, he was almost convinced that he had recognized the voice of the man who had spoken, and a momentary hope was aroused which he could not bear to lose.

Breathlessly he watched the stump behind which the man was hidden, and in a moment a form darted forth. It was crouching low and running swiftly toward a stump nearer the house. From a stump in the rear of that which had served his comrade as a shelter, the other man now also ran, and succeeded in gaining the post which his companion had abandoned.

Again, as if by a preconcerted signal, the two men advanced, both bent low over the ground, but instead of running directly for the place they were seeking, they were making a swift and zigzag course. It was evident that both understood the nature of the task before them, and were meeting it as only the backwoodsmen knew how to do.

Intensely excited by their approach, Peter had alternately watched their movements and the other side of the clearing, every moment expecting to hear the yell of the savages or the report of their guns. But thus far nothing had occurred, and at last, unable to restrain himself longer, he dropped the trap-door into its place, and leaping down the ladder, still holding his gun in his hands, rushed swiftly to the door in the rear of the cabin, and lifting the bars that held it in place, stood ready to open it for the men when they should arrive.

CHAPTER XIII

An Unexpected Ally

A SUDDEN dash, a quick, concerted movement, brought both men to the cabin, and as soon as they had entered, the door was again closed and barred. Not a shot had been fired at them, nor had a sound come from the forest where the Indians had been concealed.

Puzzled as Peter was by the fact, even that was forgotten in the excitement with which he was gazing at the two men. One of them was his friend Sam, the hunter, and the other he quickly perceived from Sarah's actions was her uncle himself. It certainly was strange, Peter thought, that these two should appear at this moment of all times, but the condition of the roof of the cabin did not admit of any delay or explanation.

"Sam," he said excitedly, "the roof's on fire!"

"So I heard ye say. S'pose I'd better get up there?"

"Yes, yes! Follow me!" exclaimed Peter, running to the foot of the ladder. It seemed strange to him that the hunter was not more excited, for to Peter the danger of the building's burning was even

133

greater than that from the besieging savages. Sam's manner was always cool and self-possessed, but his apparent lack of haste now was irritating.

Disregarding the impatience of his young friend, Sam mounted the ladder and, lifting the trap-door, gave one quick glance all about the border of the clearing.

"Humph!" was his low ejaculation; then quickly swinging himself out upon the roof he slid toward the one place where the fire appeared, and without haste or fear stamped upon it, at the same time flinging the burning ember to the ground.

The hunter's recklessness was to Peter almost unexplainable as he watched him from the trap-door. His own offer to aid had been lightly ignored, and in a brief time Sam once more crawled back to the entrance and followed Peter to the loft below. Then both descended to the room where they had left their friends.

Here they found the hunter's recent companion bending over the prostrate form of Timothy Buffum and endeavoring to discover how serious his wounds were.

He shook his head as the two returned to the room.

"Fire's out," said Sam, simply.

"Yes, I saw it wasn't much of anything. I knew ye'd have no trouble with it."

"That's as may be."

"Help me with this man. We'll put him on those skins in the corner of the room."

Together they tenderly lifted the man and placed him upon the bearskins. The hunter then carefully examined the wounds, and when that task had been completed, he rose and said to his companion: "Nothing serious, in my 'pinion. He's had a hard bump, and that's about all I can find that's wrong."

"Yes, it is serious, too!"

Peter and Sarah as well as the two men turned sharply about at the unexpected declaration and beheld Timothy sitting erect, with his face bearing an expression of vexation.

"Yes, 'tis serious, too," he repeated angrily. "Nobody ever seems t' think that I have anything the matter with me. I can be knocked through the ice and into the lake or almost have my head taken off with a tomahawk, but all everybody says is, 'Nothin' much the matter with him.'"

"Did we hurt you, Tim, when we moved you?" inquired the hunter, soberly.

"Terribly! I thought I couldn't stand it."

"What made ye, then?"

"What made me? I don't know what ye mean."

"Why didn't ye get up an' walk?"

"I couldn't."

"Why don't ye get up now?"

"I can't."

"But ye are up. At least ye're part way up. I'd finish it if I was in your place."

A groan escaped Timothy Buffum's lips as he fell back upon the bearskins. The hunter laughed and turned abruptly to see what Peter was doing. One glance had shown him that the barn was burned to the ground, and that nothing more could be done there. Within the house the pallid face of the suffering woman had required no further proof that she was in a serious condition. The children were clamoring for their breakfast, which Sarah, as soon as the fire in the roof had been extinguished and it had been made plain that there was no immediate prospect of an attack by the Indians, had been preparing. Apparently unmindful of danger, and disregarding all protests, Sam had gone out to the well and brought back two buckets filled with water. It was true he had taken his rifle with him, but this was his custom, and even when he lay down for the night, no matter where his resting-place might be, the gun was seldom placed elsewhere than close at his side.

He had insisted also upon starting a fire in the fireplace, and the warmth was almost as pleasing as the prospect of a breakfast.

"I don't just see, Sam," said Peter, when these duties had all been done and the two found themselves together by the loophole for a moment.

"Don't see what?"

"Why it is that you think the redskins are all gone."

"I don't."

"You don't?"

"No; an' what's more, Peter, there'll be dozens soon where there's one now."

"When?"

"Before another mornin'."

"What do you mean, Sam?"

"Jest what I'm tellin' ye! These redskins 'll have hundreds here afore twenty-four hours is gone."

"What then?"

"That's as may be."

"Why didn't they fire on you, Sam, when you were running for the house or were on the roof? They did at me when I went out there."

"Likely."

"But why didn't they, Sam?" insisted Peter.

"Now, Peter, look here. If you had some traps on the banks o' the Mohawk, an' ye see a good many mink tryin' for t' get in, d' ye think ye'd try to scare 'em off or drive 'em away?"

"No."

"Pre-e-cisely! An' if ye had a net strung across the river, what d' ye think ye'd do if a whole school o' fish should start for it? Would ye try for t' scatter 'em?"

"No."

"Well, it's the same way here."

"If we should try to get out, would they fire on us?"

"Jest a leetle," replied the hunter, smiling slightly.

"So this is just like an eel-pot, is it?"

"Not *jest* like it, but near enough."

"How did you happen to come here, Sam?" demanded Peter. "You were the last man I was expecting to see."

"Disapp'inted like?"

"No; I never was so glad to see any one before."

"That's kind o' ye. Well, I'd been t' Fort William Henry — "

"Did the French army come?" interrupted Peter, excitedly.

"Yes."

"What did they do? Did they get the fort?"

"Not yet."

"What do you mean by 'not yet'? Are they still there?"

"Were last night, when I left."

"What had been done? Why did you leave? How did you happen to come here?"

"Steady, lad. Not too fast. The Frenchmen had burned a few old sheds, and that's about all, up t' date, or leastwise 'twas when I left last night. What's happened since, no man knows."

"But why did you leave?"

"Because this man came there to get some help for his wife. Nobody else seemed t' want t' come, so I jest thought I'd go 'long with him."

"I shouldn't think they'd have let you go."

"'Let' me? 'Let' me? I'm thinkin' they didn't have much to do with the 'letting.' I've never seen the man yet I'd ask to 'let' me do anything I felt like doin'. I b'lieve the colonel did say as he'd be glad to have me find out all I could 'bout the Frenchmen, an' where they were, an' how many. When I go back t' th' fort I'll tell him, though I'm thinkin' he may find out afore that, an' without any help from me, too."

"Sam, do you think they'll take the fort?"

"That's as may be."

"But do you think they will?" persisted Peter.

"No man knows that, lad. They're tryin' their prettiest. As I told ye, they've come up close an' burned a few old sheds, but that doesn't hurt anybody very bad. I think if the colonel gets help an' has grit, he may be able t' hold on; but that's as may be."

"Sam, we both of us ought to be there."

"We're here, though."

"We don't have t' stay."

"Ye want t' leave this poor man an' this girl an' these children alone with that poor sick woman, do ye?"

"What do you propose, Sam?"

" I propose t' stay right here where I am now — for a spell, anyway."

" How long ? "

" That's as may be."

" And you think the Indians 'll be back here to-night ? "

" As sure 's the sun 'll set to-night."

" What 'll we be able to do against so many, for you say you think more will come ? "

" We're goin' to get help."

" How ? "

" There's a lad here what is goin' t' the fort.   If the fort (course I mean William Henry) is still standin', it 'll show they've stood off the Frenchmen, an' now can lend a hand t' others.   If it isn't so, then this lad 'll keep straight on for Fort Edward, an' jest 's soon 's he gets there he'll tell o' what's goin' on up here, an' they'll come t' help."

" And if help can't be had from either of the forts ? "

" It's *got* t' be had, an' that's all the' is to it."

" Suppose I don't succeed in getting out all right — I mean out of this house.   You know I've not been idle all the time, and you yourself said it was like getting out of an eel-pot — "

" Tell me what you've been a-doin', Peter," said the hunter, sharply.

Thus bidden, Peter briefly related the story of what had befallen him since he had been left in the hut on the shore of the lake. The hunter listened attentively until all was told, and then said, "Get some breakfast and then go up in the loft an' sleep till I call ye."

To Peter's protests the hunter would not listen, and soon the wearied youth, having hastily eaten the food which Sarah placed before him, once more mounted the ladder, flung himself upon a blanket, and was speedily asleep.

It was afternoon when he was aroused by the touch of the hunter's hand upon his arm. "Time's up, Peter," he said softly.

"What time is it?" said Peter, sleepily, sitting erect as he spoke.

"Past noon, an' ye ought to be off pretty soon. I've had ye wait till now 'cause I thought ye'd make all the better time t' have a good rest afore ye started."

"Sam, don't you think it would be better for me to stay here and for you to go back to the fort?"

The hunter smiled, but made no reply save to shake his head.

"They'd listen to you at the fort more than they will to me," continued Peter, "and if I go back, they'll be likely to make me stay. It's different with you, for you didn't enlist, you know. I'm willing to go. It isn't that —"

"No, you're the man to do it!"

Peter hesitated but a moment, and then descended

to the room below. No changes had been made there. The hunter, nodding his head in the direction of Timothy Buffum, said, "Peter, you want him to go 'long with ye?"

Peter shook his head in response, but the man suddenly arose and declared that he too was determined to depart from the cabin.

CHAPTER XIV

To the Fort

"BETTER let him go 'long with ye, Peter," said the hunter, quietly.

"He's been hurt," replied Peter, to whom the suggestion was far from being pleasing. "I'm sure you want me to get inside the fort just as soon as I can do it. It'll be hard enough, anyway, and if I have a man with me who can't do anything to help, I'll be kept back just so much the more."

"He'll keep up, Peter," said Sam, smiling grimly. "Ye needn't have any fear o' that. And if ye find he doesn't, why ye'll jest have t' leave him t' look after himself. He'll be no good here, an' worse 'n th' children."

"I'll do as you say, Sam."

"Better be off, then. Ye can get t' th' fort afore sunset if ye start now."

"My worst trouble will be right here when I start, won't it?"

"That's as may be. There's no dodgin' it, lad, that ye've got a pretty good job afore ye. But ye'll do it if anybody can, an' besides, ye jest

143

must get through, if these people here are to be
saved."

The hunter's manner impressed Peter afresh with
the danger that beset them all. That the Indians
would return at nightfall, reinforced by numbers
of their comrades, Sam firmly believed, and Peter
also shared in the belief. At the present time not
more than five were watching the little clearing.
They must know of the coming of the two men,
and that the defenders were just so much the
stronger. If some of the red men had gone to summon their comrades, the numbers of the besiegers
were reduced ; and if some also had been hit by the
bullets from Peter's rifle, as he confidently believed,
then it would be impossible for them to spare any
more to follow in pursuit of Peter and his companion, Timothy Buffum.

The hunter had suggested that when Peter
started he should run for the adjacent forest,
trusting to the suddenness of his departure from
the house and the quickness of his movements to
shield him from an attack. At the same time
Sam had promised to be ready to fire upon any
Indian who should expose himself, and was confident that he would be able in this way to check
the beginning of any pursuit, and so give his friend
a slight advantage, if afterwards he should be
followed.

The greatest uncertainty was as to the division

and arrangement of the little force of besiegers.
Up to this time apparently they had kept well
together, and had been for the most part among
the stumps or in the border of the forest beyond
the barn. From this position they had been able
to see three sides of the cabin at the same time,
and easily maintained their watch. Whether or
not some of their number had been stationed where
the fourth side of the rude little house could be
seen was a question which could be decided only
by a trial. It did not seem probable to Peter that
this side could have been left entirely unwatched,
but he did not express his fears, and began to make
his preparations for departing.

Acting upon the hunter's advice, he did not take
with him his snow-shoes, for everything that might
retard him in any way was to be discarded. There
had been a sharp freeze the night before, and it was
hoped that the ice and crusted snow would bear
his weight, and that all that he would need would
be a gun, his powder-horn, and bullet pouch. If
fortune should favor him, he would be able to
enter the fort within a few hours.

"Now bend low and make for those bushes,"
said the hunter, when at last all the preparations
had been completed and the door in the rear of the
hut was opened sufficiently to enable Peter to see
his way before him. "Tim will start at the same
time, and ye mustn't either one o' ye let the

grass grow under yer feet. An' remember Lot's wife."

"What about her?" demanded Timothy Buffum.

"Nothin' much, only jest keep her in mind, that's all."

There was a hurried good-by. Peter took one quick look about the room, and what he saw he never afterwards was able to forget. There was the sick woman in the corner and her husband standing disconsolately by her bedside; the children, silent, grave, puzzled by the strange events and unconsciously sharing in the excitement of their elders, were huddled together and silent; Sarah, her pale face lighted by a smile that somehow was wonderfully cheering to the young soldier, and Sam, quite determined, yet withal exceedingly anxious, as Peter could easily perceive, were both striving to appear more hopeful to him than either really felt.

Peter took it all in with one swift glance and then turned to the hunter, saying, "I'm ready, Sam."

"You ready, too, Tim?" inquired Sam.

"I suppose so, I don't really feel —"

"You want to stay here and take your chances, or take your chances and go?" demanded the hunter, sternly.

"I'll go."

"Then off with ye!" and as he spoke he opened the door. Instantly Peter darted into the clearing,

and exerting himself to the utmost, began to run
swiftly toward the border of the forest, not even
glancing behind him to perceive if Timothy Buffum
was following him.   He was stooping low and
bending forward as he ran, his gun held in his right
hand and his gaze fixed upon the place he was
seeking to gain.   At the same time his ears were
alert for the sound he expected and dreaded to
hear.   On and on he bounded, leaping over the
uneven ground and breathing loudly in his
excitement.

He had arrived at a point only five yards distant
from the shelter he was seeking, when the cry was
heard which indicated that his flight had been discovered.   There were yells and the report of a gun,
and a moment later the sound of another report
which Peter knew must have come from Sam's
gun.   Still he did not glance behind him.   The in-
junction of the hunter was not forgotten, but the
sight of the friendly trees was a still greater
incentive.

He gained the shelter and then, darting behind
a tree, looked back at the clearing.   Another
yell had broken in upon the stillness, and Peter
soon understood the meaning of it; for he could
see that Timothy Buffum, after having crossed half
the clearing, had stopped, and, standing irresolute
for a moment, was apparently tempted to return
to the shelter of the cabin.

"Come!" called Peter, and then he stopped abruptly, aware that he could not help the man, and that he only revealed his own presence by his shout. Yet he was watching Timothy with an anxiety he could not express. What had come over the man that he should falter now? Had he been hit by a bullet? Even while the questions were rising in his mind, he saw Sam step forth from the door, and, raising his gun to his shoulder, fire in the direction of the opposite side of the clearing. Then, calling to Timothy, the hunter started swiftly toward him.

Before he had taken a half-dozen steps, however, the man seemed to recover in a measure from his confusion and began to move toward the house, whereupon the hunter darted back to the open door. Peter watched both for an instant, and then, realizing that Timothy was not to be with him in the attempt to go to Fort William Henry, recalled to his own peril he at once turned about and began to run swiftly through the forest. He heard another wild cry behind him, but was not able to determine what it meant. He did not even know whether Timothy Buffum had been hit or not, or if he had been able to regain the protection of the cabin.

Aware, as he was, of the direction in which Fort William Henry lay, his own purpose now was to reach it in the shortest possible time. There

arose in his mind the hunter's suggestion that the
fort itself might have been burned and the men
made prisoners. What could the little force of
three hundred and fifty men do against the division
of the French army, consisting of sixteen hundred
men, sent by the Canadian governor for the purpose of destroying or taking the fort of the English
settlers? A new fear was in Peter's heart now as
he ran forward, and the thought of being compelled
to go on to Fort Edward for aid was strong upon
him.

Satisfied, after he had run until his breathing
was becoming labored, that there was no open
pursuit, although he understood the nature of his
foes too well to permit himself to believe for a moment that he had escaped all peril of that nature,
he began to slacken his speed and assumed the
slower, steady run which the Indians used and
was frequently employed by the younger men
among the settlers when they were making long
journeys on foot.

Not once did he stop for rest. The sun was low
in the western sky when at last he arrived at a
place which he recognized as being not more than
a mile distant from Fort William Henry itself.
And not a sign of pursuit had thus far been seen.

His problem had now changed, and he must be on
his guard against the Frenchmen or Indians who
might have surrounded the fort. Just as the sun

disappeared from sight, he halted and prepared to move forward more slowly now and with increased caution.

Suddenly he heard from before him the sound of guns. Faint and far away the sounds arose, but there was no mistaking them: they were the reports of rifles or muskets. Startling as they were, Peter's heart was rejoiced at the sound, for it implied that the fort had not as yet fallen.

He moved forward cautiously, and soon noticed that the sounds could be heard from more than one place. And the firing, although the sound of it did not come at regular intervals, was evidently not without some systematic direction.

As Peter advanced he found that the shots he still continued to hear came from what seemed like a huge circle, and at once he understood. Fort William Henry was still standing, and the invading army, having surrounded it, were firing upon it from all sides at the same time. In his excitement even the plight of his friends for whom he had come to seek aid was forgotten. If the fort was entirely surrounded, how would he be able to break through the lines and gain an entrance? The question was puzzling. His first impulse, to seek the shore of the lake and approach the fort on the ice, was abandoned when he discovered that the sounds came from afar as well as from the near-by places, and he concluded that the Frenchmen were as busy

on the frozen lake as on the snow-clad shore. He
feared to advance, although he dared not remain
where he was. Alternately moving forward and
halting to listen to the sounds which became every
moment more distinct, he suddenly heard the thunder of cannon, and knew that the men in Fort
William Henry were either hard pressed or were
firing upon bodies of men that could be distinctly
seen. He could hear no balls crashing their way
through the forest, and soon concluded that the fort
must be in greatest danger from the lake side, and
that the men were firing in that direction alone.
The sound of the muskets and rifles before him still
continued, however, and it was plain that his own
peril was not lessened.

Suddenly, as if by some preconcerted signal, all
firing ceased, save that of the cannon from the fort,
and soon even that too was no longer heard. A
silence intense and even more fearful than the reports of the guns had been, rested over all. For a
brief time Peter listened, peering keenly into the sombre forest before him, but was unable to discover the
cause of the strange stillness. Every tree seemed
to be the hiding-place of some foe. The leafless
branches that swayed above him were like long
arms stretched forth to grasp him. A fear such as
he had never felt before swept over him, and he
was almost tempted to turn and flee from the
region. To know that somewhere between the

place where he stood and the little fort which he
was seeking, were men armed and eager to seize
him, and that if he should move forward he might
at any moment be in their power, was almost more
than he could endure.

His suspense, the terror inspired by the intense
stillness and loneliness, suddenly departed when he
became positive that he could discern the forms of
men moving amongst the trees before him. There
were a half-dozen, at least, in the little group, and
these were speedily followed by others. They gave
no heed to the locality where Peter was standing,
evidently believing that no danger was to be feared
from that side, but pushed steadily forward, moving swiftly, but without any semblance of order, all
in the same direction.

For a long time the scattered men passed in front
of the place where Peter crouched behind a tree
watching them, and then no more appeared. He
waited for what seemed to him to be hours, but
not another man was seen. Convinced that all had
been withdrawn, although he could not conjecture
why, he too began to creep stealthily forward, determined once more to attempt to make his way
into the fort.

A great blaze in the sky, the distant calls and
cries of men, the fact that the forest became all
light before him, caused him to change his plan,
and he turned to his left and began to run toward
the shore of the lake.

CHAPTER XV

IN THE FORT

AS Peter drew near he could see that bodies of men were running along the shore and out upon the lake. Two sloops, held fast in the ice not far out from the shore, were burning, and a large number of bateaux that had been drawn from the water in the preceding autumn, when the cold weather had arrived, were all in flames.

Peter stopped for a moment, watching the fires and listening to the roar of the flames as well as to the faint shouts of the distant men. The tongues of fire leaped up the masts of the sloops and darted from the bows. It was a sight to be remembered, and trembling with fear and anger, he stood almost spellbound for a time and watched the destruction of that which had cost the colonials so many days of hard labor. It was maddening, and yet he was powerless to aid. A half-dozen men came rushing past the place where he was standing, and he drew back hastily among the shadows; but no heed was given him, and the men soon disappeared from sight.

A shout from the direction of the lake caused him

once more to step forth from his hiding-place, and
soon he discovered the cause of the renewed commotion. A band of men were advancing swiftly
from the fort in the direction of the burning vessels.
Again the shouts rang out and the exultant cries
were redoubled. But the men from the fort kept
steadily on their way until they had approached
near enough to convince themselves that any attempt to save the blazing sloops was useless; then
in all haste and in apparent disorder they turned
and fled back toward the shelter of Fort William
Henry.

A sudden determination swept over Peter, and
he decided to try to join the running men and
with them gain an entrance into the fort. He
darted from his hiding-place and ran swiftly toward
the shore. The men were not far away, and in his
eagerness to join them he put forth all his efforts,
and was soon in their midst. No attention was
paid him, and in a brief time all had regained the
fort and been received within.

The excitement among the garrison was great
until it was perceived that it was impossible to
save the burning sloops and bateaux, and that there
was slight immediate peril for the fort itself and
its inmates. While it lasted, Peter knew it was
impossible even to ask for the aid for which he
had come; but when a measure of order had been
restored, he at once sought out the quarters of

Major Eyre, determined to make his plea, and if possible obtain permission for a few men to return with him to rescue his friends from the perilous situation in which he had left them.

He speedily discovered, however, that it was impossible for him even to be admitted into the major's presence at that time, and baffled and keenly disappointed, he sought quarters for the night. Special guards had been stationed and the watches doubled, for the fear of a night attack rested upon the entire garrison. No special duty had been assigned to Peter, and realizing as he did the uselessness of attempting anything more before morning, he was soon asleep. When the sun rose on the following day, he had obtained a greater rest than he had known for some time.

It was Sunday morning. Within the fort preparations were being made for the religious services of the day. The chaplains were earnest men, and the serious nature of the work before the soldiers made all thoughtful and responsive. The fort itself was still safe, and the threatened attacks of the invading force had not materialized.

Although the Frenchmen and Indians greatly outnumbered the defenders of Fort William Henry, their method of making war, particularly on the part of the Indians, was not of an aggressive nature. It was rather an advance in the darkness, or the cutting off of supplies or of parties sent to aid, and thus far their efforts had proved unavailing. But no one knew what a day might bring forth, and the knowledge of their enemies' usual method of procedure kept every one in the garrison alert and watchful.

Just as the men were assembling to listen to the words of Chaplain Williams, a cry arose that the enemy was approaching. The service was abandoned, and all the men rushed to places on the walls of the fort, from which the advance of the allied forces could be seen.

The rumor was soon proved true. Peter could see the long lines of the men as they filed out from the forest and moved across the lake. They carried scaling-ladders, and it was evident that Rigaud was moving in a manner to impress the defenders of the fort with his own numbers and power.

" Not much to fear from them to-day."

Peter turned sharply as the words were spoken and saw Chaplain Williams standing close behind him.

" Why not ? " he inquired.

" Because that's not their way when they have anything to do. If we had heard of their coming in the darkness, or they had tried to conceal themselves or hide their approach, we might well begin to fear. But as it is, I am confident this is all for display."

Peter was silent as he watched the movements of the men. The chaplain's surmise seemed to be correct, for the troops soon halted at a safe distance from the guns of the fort, and then several of them could be seen advancing from the line and ostentatiously waving a red flag which they carried.

"They want a parley," said the chaplain.

"There'll be no attack now?"

"Probably not."

"Then this is the time for me to see Major Eyre. He wouldn't let me last night. I left some friends a few miles back here, and they're in desperate trouble. The redskins are all around them and have burned up their barn, and I got away so that I could come on here to the fort and get some men to go out and help them. I ought to have gone back last night! I must see the major! And if there isn't any likelihood of an attack being made here now, then he'll see me. Won't you go with me? I think he'd be willing to see me if I should come with you," pleaded Peter, doubly eager because of the long delay he had been compelled to suffer.

The chaplain shook his head dubiously. Both could see that an officer with a few men had already left Fort William Henry and was advancing to meet the Frenchmen. "I'm afraid just now—" the chaplain began.

" Please come ! Please help me ! " interrupted Peter. " There's a sick woman and little children, and — "

" I'll try it," said the chaplain, quickly. " I have but slight hope; but we can at least make the attempt."

Together the two started for the quarters of Major Eyre. When they arrived at the room he occupied, the chaplain left Peter outside, while he himself entered. Left alone, Peter's fears and anxieties swept over him in full force. He could see again the little cabin and its inmates, and hear Sarah's firm words as she busied herself in the task of caring for the sick. By this time, doubtless, the numbers of the besiegers had greatly increased, and it might even be that the house had been burned and its inmates had fallen before the bullets or tomahawks of the savages. Peter's confidence in Sam was boundless, and he knew that all that could be done he would do, but it was hardly possible that even Sam's many resources would avail to hold the little place.

And they had been depending upon him to bring aid ! Peter blamed himself for the delay, although he knew the charge could not justly be made against him. He had gained an entrance into the fort and had done his utmost to obtain an interview with Major Eyre, but thus far had failed. The burning sloops, the blazing bateaux, the pres-

ence of the Frenchmen and Indians in such numbers before the walls, had been more than sufficient
to demand all the time and energy of the hardy
leader, and all other matters must be kept in the
background. Now as Peter thought of the entire
invading army which had halted within plain view
of the defenders of the fort, his heart sank within
him at the faint prospect of his obtaining the aid
for which he had come. He could hardly expect
even an interview. His somewhat bitter reflections
were interrupted by the return of Chaplain Williams.
"Come," he said quietly; and Peter's heart gave a
great throb as he followed him into the room of
Major Eyre.

The major was standing when they entered, and
as the chaplain said, "This is the man," Peter
began to tell his story. He spoke rapidly, his
eagerness revealing itself in the very tones of his
voice, as he briefly related the story of the defence
of the cabin and the desperate plight in which its
inmates now were.

The major listened thoughtfully, but when Peter
stopped and looked up anxiously for his reply,
he said, "I'm sorry, my man, but it can't be
done."

The expression of disappointment and despair
on Peter's face perhaps moved him to explain, for
he said kindly : "I know it seems hard, but not a
man can be spared from Fort William Henry now.

We need four times as many as we have, and it is utterly impossible to think of letting even one go."

"But I can go back?" said Peter, aghast.

"You have enlisted?"

"Yes, sir.   But —"

"Then you cannot go."

"There are women there, and children —" began Peter, bitterly.

"It's sad.   It's hard.   I understand all that, my man.   But you can readily see that if we don't hold this fort, it will not be one cabin or one family that will have to suffer, but hundreds of them.   If these savages once get an open road to Albany, I dare not think of what would follow.   For the sake of hundreds of women and children we must hold this place and keep them back.   Not one man can leave us.   Perhaps in a few days something can be done," he added sympathetically.   "I know Sam, and I don't think he'll let the place go easily. It may turn out better than you fear.   And the best way to help your friends will be to make your strongest fight right here.   I'll —"

The major stopped abruptly as a half-dozen officers together entered, having in their midst a man, evidently a French officer, blindfolded.   It was impossible for Peter or the chaplain to leave the room ; and, pushed back against the wall, they watched the proceedings, for the time almost unmindful of what it was that had brought them there.

"What's this? Who is this?" said the major in a low voice to one of the officers.

"Le Mercier, the commander of the Canadian artillery," replied the man. "He said he had a message for you from Rigaud. He came with a red flag, and we went out to meet him, blindfolded him, and brought him to you."

"Take away the bandage," said the major, quietly.

He was obeyed, and the French officer bowed graciously. "I congratulate you, Major," he said in English, "upon the excellent defence you have made."

"I thank you," replied Major Eyre, bowing slightly. "Our defence is to be measured by the daring and ability of our enemies."

"It is gracious of you to say so," responded the Frenchman, bowing in turn. "It is because of this mutual understanding that I am come."

Major Eyre made no response, but the question he would ask was apparent in the expression of his face.

"I am come," resumed Le Mercier, "to invite you to give up the fort peaceably, and so save all unnecessary and further bloodshed. Doubtless you are aware that we outnumber you four or five to one. We have equipments of the best, supplies, and, as you yourself are pleased to say, a spirit of daring and determination among our men that cannot easily be held back.   We will grant the best of terms and such as men of honor can accept without the slightest fear of reproach."

Major Eyre still was silent.

The tone of the Frenchman's voice slightly changed as, after waiting for the major to speak, he resumed : " I offer you this simply from my desire to save you an unnecessary loss of men.   We are not bloodthirsty.   We have no desire to inflict useless suffering.   But this fort we must and shall have.   In case it is not given up peaceably, you yourself know well how difficult it will be for us to hold back our Indian allies, if once it falls into their hands.   It is to save you all from this that I am come."

Major Eyre's eyes flashed as he said in reply : " You have praised our defence, but let me tell you it is as nothing compared with that which will be made.   There is only one way in which you can obtain possession of Fort William Henry."

" And that is ? "

" To take it."

" That we can do easily, but it will not be so easy to prevent the massacre that will certainly follow."

" That is for you to determine; but first you

must take the fort. As long as one man is left to defend it, it will be defended."

" And that is your final reply ? "

" It is my only reply."

The Frenchman bowed, the bandage was replaced over his eyes, and he was led from the room.

CHAPTER XVI

THE ATTACK

PETER followed the company as it departed from Major Eyre's room, for he knew that any further request for aid would be fruitless. The blindfolded Frenchman was led to the shore and over the ice to the place where his comrades awaited him; and as soon as his conductors had returned to the fort great excitement arose in the garrison, for it was seen that the French were preparing to advance and storm the works. Even Peter Van de Bogert, anxious for the people in the little cabin he had left the preceding afternoon, and chagrined as he was at the thought of the interpretation which Sarah and Sam might put upon his failure to return with help, was now so strongly aroused by the sight of the advancing French army that he was as eager as his comrades to defend the fort against the invaders.

The entire force was in motion, and as soon as it had approached within range began to fire upon the walls. No response was made by the garrison to the harmless fusillade, for orders had been given for the fire to be reserved until the enemy was near

164 .

enough to receive its full effect. The excitement was intense, and men and officers alike were eagerly waiting for the supreme moment to arrive.

Suddenly the advancing force halted, then turned sharply to the left and soon disappeared from sight within the wooded shores. At first the defenders of Fort William Henry hardly dared to trust the evidence of their own eyes. It seemed well-nigh impossible to believe that after all the parade and the bold declarations of Le Mercier, the attempt to storm the works would be abandoned without one blow being struck. And yet the Frenchmen and their Indian allies had certainly abandoned the effort, and cheer after cheer rose from the little force of defenders as their enemies disappeared.

"I fear this rejoicing is not well founded."

Peter turned sharply as he heard the words, and beheld the chaplain by his side. "We can rejoice that they are gone for this time, anyway," he said eagerly.

"'And sufficient unto the day is the evil thereof?'" inquired the chaplain, smiling as he spoke.

"There's good authority for that."

"So there is, so there is, and I would be the last man to deny it. But I am fearful that the fort has not seen the last of them yet."

"The major says I must stay here," said Peter, abruptly.

"Yes, I heard him."

"I don't think I shall."

"Yes, you will, my son."

"If you were in my place, you'd try to go back to your friends," rejoined Peter, bitterly. "I can see Sarah and her poor sick aunt and the little children and Sam, even now, and it doesn't seem to me I can endure it to stay away."

"You have done your best, and no man could do more."

"Yes, sir, I think he could. He could go back himself."

"Against the major's orders?"

"He hasn't any right to make such an order."

"He evidently thinks he has," said the chaplain, quietly.

"But he doesn't know. He doesn't understand," protested Peter.

"Now, my son," began the chaplain, calmly, "you must look at things as they are; not as you would like to have them. You left the cabin, and I have no doubt your friend did not really think you would succeed —"

"I promised her," interrupted Peter, sharply.

"Her? I thought 'twas the hunter, Sam, who sent you for help."

"He did," replied Peter, in some confusion.

"By this time either he has escaped or driven off his adversaries. In either event, you could not avail in any way if you did return. You lived up to your promise. You succeeded in entering the fort, and made known your word to the major. That no aid should be sent or that you yourself should be compelled to remain here is no fault of yours. As all good soldiers, you must obey —"

"But the Frenchmen have gone," suggested Peter, bitterly, "and now there is no need of my being here. I must go, Dominie! If you only knew —"

"I know you will obey and strive to be patient."

Peter shook his head, but made no other response. It was useless to talk further of a matter which no one except himself understood. It was all well enough to require him to stay in the fort when the enemy was in plain sight and advancing upon it; but to refuse him permission to go back to the aid of his friends when the threatening danger had passed, was more, he thought, than he was able to endure, and more than should be expected of him.

All through the afternoon he deliberated with himself concerning the matter. There was something in the chaplain's suggestion that by this time a crisis must have come to the people in the cabin, and either the peril had passed or — Peter shuddered as the alternative presented itself and resolved to seek Major Eyre again and once more lay before him the exact condition of affairs and beg at least for permission to return, if no aid could be given.

Filled with the thought, he sought the major's

quarters, but was bitterly disappointed when he was informed that he could not be admitted. Night fell, and still Peter struggled restlessly with himself, at one moment determined to escape from the fort, come what might, and go back to his friends in the cabin, then again striving to reconcile himself to the chaplain's words, persuading himself that whatever was to have befallen the hardly beset party had already occurred, and that it was only the part of wisdom to accept his place in the fort and do his best there. He had given a promise there also, and was not his word as binding in one place as in the other?

Feverish, unable to close his eyes, Peter at last rose from his place among his sleeping companions and went out into the open air. As he did so there was a scurrying among the guard, and he heard the low, sharp call of some of the officers. Instantly alert, he discovered the cause of the confusion, and in a brief time the entire garrison was called out, for the French, under cover of the darkness, were again moving upon the fort. A crisis far more serious than that which had been faced in the morning was now at hand, and Peter was as ready as any of his comrades. Every man held his gun in readiness for instant use, and stood trying to pierce the darkness, and listening to the sounds of the approaching foe. The men had been commanded to fire in the direction of the sounds, without wait-

ing for orders, and the sharp reports of the rifles
were almost continuous. Peter, too, was busily
engaged in firing into the darkness.

But the attack against the fort was only a pre-
tence, for the real purpose of the advancing force
was to set fire to the buildings outside. A number
of storehouses, a sawmill, a hospital, the huts of
the rangers, and a sloop on the stocks were not
far distant, besides great piles of cord-wood and
planks.

Under cover of the darkness, and while their companions were making the pretended attack upon
the fort itself, numbers of the Frenchmen and Indians, with fagots and resinous knots in their hands,
crept stealthily up against the side of the buildings
farthest from the fort, set them on fire, and before
the flames burst forth had again escaped into the
darkness. It was not long before a great blaze
seemed almost to surround the fort and its sadly
beset little garrison. Showers of burning embers
were carried by the breeze into the fort itself. To
attempt to go forth and to subdue the fires would
be but to expose the men to the bullets of the
enemy, who were themselves safely sheltered by
the outer darkness.

Besides, the garrison was exerting itself to the
utmost to protect the barracks from the showers
of sparks that fell almost steadily upon it. Again
and again a flame would start up in some building,

and by the time it was extinguished, another would demand all their efforts. All night long the desperate men labored to prevent a calamity more to be feared than that of falling into the hands of Rigaud. Even after the light of morning appeared, the desperate conflict continued; but at ten o'clock the wearied men found relief. The fires had been gradually burning low, though the showers of sparks were still to be fought, when the gray clouds broke and snow began to fall. Thick, heavy, moist, the flakes soon accomplished what the men had been unable to do, and the peril was past. So thick was the snowfall that the watchers could see only for a distance of a few yards from the fort. All day long and throughout the following night it continued; and when at last the storm ceased, the damp snow covered the frozen earth and the ice on the lake to the depth of three feet or more.

The fear of an attack during the storm had been strong, but the Frenchmen lay hidden in their camps until just before dawn on Tuesday morning. Then a little band of twenty started forth to set fire to a sloop, some storehouses and other buildings, and several hundred scows and whale-boats, — nearly all that had thus far escaped the torch.

In the dim light the party was seen, and instantly there was a commotion within the fort. The excitement under which the men had been laboring, and then their long watching for the com-

ing of the enemy, now found its opportunity to display itself in the very kind of action for which
they were longing.

There was a hurried call for volunteers to drive
back the bold band which had come so near to the
fort, and so prompt was the response that the difficulty was not to obtain men but to select them.
Peter Van de Bogert, restless and anxious, considered himself fortunate to be included in the
number of those who donned snow-shoes and
sallied from the fort to protect the boats and
buildings and drive back the Frenchmen.

So active had the latter been and so stealthily
had they crept up that their presence had not been
discovered until they were close to the places they
were seeking. As the men from the fort approached they could see that already fires were
burning, and with a shout they increased their
speed. An answering shout of defiance greeted
them, there was a hurried movement among
the Frenchmen, the blazing fagots were thrown
hastily among the boats; and then the little force,
consisting of only a score, started swiftly back
toward their camp. The colonials were instantly
divided into two bands, one of which rushed to
the place where the fires were burning, while the
other started after the fleeing Frenchmen who, like
their pursuers, were equipped with snow-shoes.

At the word of the leader the colonials fired, and

a shout arose when it was seen that several of the fugitives had fallen. Peter was in the pursuing division and had fired with the others when the command was given, but though he was not able to determine the effect of his own shot, his heart was filled with a wild rage as he gazed at the men before him. There was no recognition of their bravery, no thought of mercy. These were the men who had burned buildings, and almost succeeded in destroying Fort William Henry itself. They were the enemies of all good, and the wild rage of war consumed him as he ran on with his comrades in the hot pursuit.

Again the party fired, and again men were seen to fall while the rest of the band rushed forward with increasing speed. There was no attempt on the part of the Frenchmen to return the fire, their every effort being centred upon reaching their camp safely. Once more the fire was opened with deadly aim. Until now it was evident that only about half who had advanced would succeed in returning to camp; but this half redoubled their efforts and were soon beyond the chance of being overtaken. Then the word was given for the colonials to return to the fort, a certain number being delegated to care for the wounded, and the remainder hastening to the assistance of their comrades who were fighting the fires.

Only partial success had rewarded the efforts of

the French this time. The sloop was burned as well as a few small buildings, but all else was saved by the energy of the defenders.

On Wednesday morning the sun rose glorious and strong. The snow glistened like crystal, the air was warmer with the breath of spring; but none of these things impressed the men in Fort William Henry, for they were watching a scene far out on the frozen waters of the lake. Rigaud's men were retreating, and the present attempt to destroy the fort and capture its garrison had failed.

Great was the rejoicing among the hardy colonials, but Peter Van de Bogert's thoughts were speedily directed into another channel that more nearly concerned him and the friends whom he had left in the little cabin.

CHAPTER XVII

The Return

SOON after noon on the day when the army of sixteen hundred Frenchmen and Indians had abandoned this attempt to take Fort William Henry and had withdrawn from the region, Peter received a summons from Major Eyre which he was not slow to obey. Reporting at once to the major's room, he found that officer in exceeding good humor and most cordial in his greeting.

"Glad to see you, young man," said the major. "I think now we can do something for you."

"If it is not too late," replied Peter, soberly.

"Your friend Sam has been able to take care of himself; you need have no fear as to that. Patience is the greatest of all virtues. If we had been frightened here and had not hung on, the fort would have fallen. As it is, you see we are still doing more than could rightly have been expected of us, for without any help from Fort Edward we've held the fort and driven the rascals away. If they had made a rush on us, they would surely have overwhelmed us, for they outnumbered us

174

five to one. And you'll find the same thing true
in the cabin, for I'm going to let you go back now
and send a small party with you."

Elated as Peter was by the information, he was
not able to share in the confidence which the
major appeared to feel. Knowing, as he did, the
condition of affairs in the little cabin when he had
left, and also aware of the fear in the minds of the
men in the fort during the past few days, he was
by no means so confident as the officer appeared to
be that the threatening peril was over.

"Do you really think the Frenchmen have
gone?" he ventured to inquire.

"I know they have. My scouts have followed
them and have brought me word."

"And you don't think they'll come back?"

"Not for the present, and before they can get
ready to try it again General Webb will be ready
to move on them. No; we have not much to fear
from another attack just now."

By no means convinced, for his own experience
and knowledge of Indian ways had made him fearful, Peter was nevertheless so rejoiced over the
permission to return to his friends and the promise
of aid that he was already impatient to be gone.
His uneasiness was apparent to the major, who
smiled as he said : —

"You're to go at once, if you so desire. I have
arranged with Jeremiah Stubbs to take five of his

men and go with you, and he's waiting for you now, I doubt not."

Peter's face lighted up instantly. He knew Jeremiah Stubbs, one of the sturdiest of the "rangers" that had followed Captain Rogers from New Hampshire. Tall and powerful, the man was so serious in all his undertakings that he was looked upon, even by those who knew him best, as somewhat peculiar. A firm believer in the efficacy of "signs," ever ready to account for any unusual occurrence by the aid of supernatural forces, he nevertheless was a man to be thoroughly depended upon and implicitly trusted. It was a common remark in the army that no one had as yet ever seen a smile on the countenance of the worthy Jeremiah ; and as he was now a man in middle life, there was but slight hope that he would ever display one. His duties as a scout and a follower of Rogers had been of the most perilous nature ; but he had not drawn back from any call, and already had had almost as many exciting adventures as his doughty leader. So when Peter Van de Bogert heard from Major Eyre that Jeremiah Stubbs was to be the leader of the little force that was to accompany him on his return to the cabin, his feeling of confidence became stronger, although his anxiety as to the fate of his friends was in no wise abated.

"I trust you will find everything as it should

be," said the major, as Peter prepared to depart.
"I have no doubt you will, and I should not be
surprised even to hear that you met your friends
on their way to the fort. Many are coming here
for the protection they need."

The confidence and elation of Major Eyre were
inspiring, and even Peter could not prevent himself
from sharing his feelings in a measure, though his
fear as to the fate of the people in the cabin had not
wholly departed. He bade the major good-by, and
started for the place where he was to meet Jeremiah Stubbs and his men. That worthy leader
gave him solemn greeting, but Peter's thoughts
were instantly centred in one young man who was
in the company. Hesitating for a moment before
he addressed him, he said : —

"John ? Is that you, John Rogerson ? "

"That's the name by which some folks call
me."

"Where did you come from ? "

"From roaming up and down the earth, like a
certain person who seems to have a good deal to
do with our worthy leader Jeremiah."

"And are you going with us ? "

"That's what they say."

"Great!" exclaimed Peter, enthusiastically. "It's
almost like seeing some one come back from the
grave to get a sight of you ! "

"Not quite so bad as that," laughed John.

" 'Tisn't two years since we were in this same part
o' the country before,[1] is it ? "

" I don't know.   It seems like ages.   I'm glad
you're to go with us," said Peter, simply.

" I'm glad you're glad.   And if I'm glad and
you're glad, too, why we'll all be glad together,
except Jeremiah, and he can't smile without making you think he has a pain in some vital organ.
Jeremiah, did you ever laugh ? " he added, abruptly
turning, with the freedom of the times, to the leader
himself as he spoke.

" I'm not one o' those what think life is somethin' to grin at," replied Jeremiah, solemnly.   " Are
ye all ready ? " he demanded.   " 'Cause if ye be, we
don't want t' waste any more time," and he glanced
at John as he spoke, as if he would reprove him.

That young man, of nearly the same age as Peter,
stalwart, rough, strong, and light-hearted, could not
be repressed entirely, for he inquired blithely, " Are
the signs all right, Jeremiah ? "

" You never mind th' signs !   I'll look after them
and you too ! "

" I'll do as much for you sometime, Jeremiah."

The leader, however, did not deign to respond
to the irrepressible youth, and the few preparations
for departure were hastily completed.   All the men
were equipped with snow-shoes and armed with
rifles.   Their clothing was mostly of deerskin, and

---

[1] See " With Flintlock and Fife."

their bronzed and rugged faces were aglow with determination. They were aware of the nature of the undertaking into which they were going, and were men who would do their utmost to aid the settlers, cut off from help and left to the terrible task of defending themselves against an enemy whose presence was feared far more than that of the prowling wild beasts.

In single file the little band departed from Fort William Henry and soon was lost to sight in the depths of the forest. The sun had been shining brightly all the morning, and the deep snow was soft under their feet. A great mass fell from one of the bending branches and struck Jeremiah on the back of his neck, whereat John Rogerson laughed aloud.

" Did you do that ? " demanded Jeremiah, angrily.

" Did I do what ? "

" Throw that snow at me ? "

" Jeremiah, it grieves me to think you could suspect me of such a thing. That snow fell from a tree and was given you for a sign."

" A sign of what ? "

" A sign that you ought not to get hit in the back."

" I never was hit in the back in my life," retorted Jeremiah, sharply.

" Move on up in front there ! " called some one in the rear, and the march was resumed.

The walking became more difficult with every passing hour. Sometimes the snow-shoes were abandoned and the men struggled forward through the soft slush that in places came almost to their waists. Perspiration streamed from their faces, and they frequently stopped for rest. The sole source of comfort was that not a sign of Frenchman or Indian had been seen, and this confirmed the belief that the entire body had withdrawn to their own fort at Ticonderoga.

It was near nightfall when at last the men, wearied by their long and difficult march through the forest, arrived at a place near the little clearing which they were seeking. Peter's eagerness to push forward now increased, and he began to run in advance of the entire line.

"Here!" called Jeremiah, sharply. "Wait a minute, will ye?"

Peter reluctantly halted and joined his companions whom the leader had called together, and a consultation was held.

"What are we waiting for?" he demanded impatiently.

"Signs," replied John, demurely.

"Ye may laugh if ye want t', but that's jest what we're doin'," said Jeremiah. "Do ye want to rush right out there in that clearin' afore ye know who's who an' what's what, an' jest let th' first redskin that sees ye draw a bead on ye?"

" It 'll be dark soon," suggested Peter.

" There's worse things 'n th' dark, 'cordin' t' my way o' thinkin'," said Jeremiah, soberly. " John, here, c'n laugh if he wants to, but we've jest got t' look up th' signs afore we go ahead a step farther. Who knows whether th' cabin's still there or not? Who can tell me if there's anybody in th' house? How d' we know a whole lot o' Frenchmen and redskins didn't stop here, and may be here now for all we know? "

There was a general assent to his words, and Peter asked quickly : —

" What shall we do, Jeremiah? What do you suggest? "

" I suggest that we two go on and all the others stay right here where they be till you 'n' I come back an' report what's goin' on up ahead."

" Are you and I to go? " inquired Peter, eagerly.

" That's what I said."

" Then come on! Let's start! " said Peter, impatiently.

" Hold yer horses! Let's get this all fixed beforehand, an' we'll know what t' do 'f anything happens. Peter 'n' I'll creep up ahead an' have a peep at th' clearin' an' likewise at the cabin. If everything 'pears t' be all right, we'll come back 'n' report; but if anything happens t' us, you'll prob'ly know it when ye hear th' guns; an' ye'll know what t' do, too. Now, then, Peter, we'll start."

Together the two men went forward and cautiously made their way toward the little clearing, which Peter had declared was but a short distance before them. The sun was low now, and there was a chill in the air, but Peter was scarcely aware of it, so excited and eager was he. If his friends were still there, he was wondering how he was to explain fully the failure on his part to bring aid before. He had done his utmost, but would Sam believe it? And would Sarah be satisfied with his explanation? And then the fear returned that neither Sam nor Sarah might be there. He unconsciously quickened his pace, but a low word from Jeremiah checked him, and he drew back to his companion's side.

They were near the border now, and between them and the open clearing only one more little stretch must be crossed. Together they crouched low and ran swiftly to the clump of cedars already well known by Peter.

A sigh of relief escaped his lips as he beheld the little log-house still standing and apparently having suffered no harm. "It's there! It seems to be all right!" he eagerly whispered to Jeremiah.

"See any smoke?" inquired Jeremiah, in a low voice.

Peter gazed intently at the rude chimney, but could not perceive the slightest indication of rising smoke. What did it mean? Was the place aban-

doned, and had the inmates fled? Or had worse
evil befallen them? A nameless terror possessed
him. He was trembling in every limb as he vainly
looked for some sign of life. But stillness rested
over all. In the gathering gloom the very cabin
itself seemed to be holding back a secret that he
longed and yet dreaded to learn.

"See anything?" whispered Jeremiah.

"Not a thing. What shall we do?"

"Go back and report, then all come up together."

"Shan't I call? I might make them hear if any
one is inside the house."

"Yes, try it," said Jeremiah, abruptly.

And Peter lifted up his voice and called; but
before he could repeat the hail, the perplexing
question was solved, and both men had learned
that for which they had come.

CHAPTER XVIII

A Difficult Journey

IT was Sarah herself who stood in the open doorway looking out in the direction from which the unexpected hail had come. To all appearances she was unharmed and well, and as Peter beheld her he turned quickly to his companion, saying: —

"You bring up the men, Jeremiah. I'm going to the house."

Without waiting for a response, he dashed across the clearing to the doorway in which Sarah was still standing. She recognized him before he gained the house, and partly closing the door, called:

"Is any one after you, Peter? Are you being chased?"

"No, no," responded Peter, breathlessly, as he dashed into the house. "Friends are out there, Sarah, and you are all safe now. I've come at last."

"Where are they?"

"Just beyond the clearing. They'll be here in a minute. Is everything all right here? Where's Sam? Has any one been hurt? How is your aunt?"

184

Before the girl could reply the recent companions of Peter appeared in the clearing and drew near the house. Both Peter and Sarah stood in the doorway waiting for them, and neither spoke until Jeremiah Stubbs, who was the first of the little band to enter, looked solemnly at the excited girl and said : —

"I hope I see ye well."

"You do, sir," replied Sarah. "I'm thankful to say we have all been spared."

The other men now came crowding into the house, glancing curiously at the inmates and displaying the evidences of their good-will in their rough and boisterous manner. Peter saw that Sarah's aunt was standing to greet the newcomers, and concluded that she, at least, was none the worse for the recent siege. Her husband, also, was there, and the children and Timothy Buffum, but as yet he had not seen anything of his friend Sam.

After the first greetings had been given and the story of the march told, the tallow dips were lighted and a homely meal was provided for the wearied men. Peter was eager for a word with Sarah, but she was too busy for any extended conversation. He did, however, contrive to ask her one question as she passed the bench on which he was seated.

"Where's Sam?" he inquired.

"Gone."

"'Gone'? Gone where? When did he go?"

" Early this morning."

There was no opportunity for further questions, and Peter strove to content himself until the supper had been eaten.   Then while all gathered about the huge stone fireplace, on which great logs were blazing, Sarah and her uncle briefly told the story of their recent experiences.

After Peter's departure the hunter had assumed charge of the defence of the lonely home.   In the first night the Indians, reënforced by the arrival of a half-dozen warriors, had endeavored to set fire to the cabin, but Sam's watchfulness and his ever ready rifle had prevented them from carrying out their purpose.   On the day following, not a sign of their presence had been discovered, and then it was decided that they had abandoned the siege. The hunter had delayed his own departure until the morning of the present day, when he had declared that he must go.   Sarah added that she herself was convinced that Sam had been fearful of some evil having overtaken Peter, and that his failure to return had caused the hunter to set forth to obtain some knowledge of him.

The reference to Peter's failure to return with aid provided the opportunity for which he was waiting, and he briefly related what had occurred at the fort, and told how Major Eyre had forbidden him to leave until the French army had departed.

"ALL GATHERED ABOUT THE HUGE STONE FIREPLACE."

The sense of security which now prevailed in the little home, the presence of the men as well as the improvement in the condition of the sick woman, all united to make the entire company far more happy than any of them had been for days.

"It isn't safe for you to stay here alone," said Jeremiah, at last.

"I'm afraid we'll have to," responded Sarah's uncle.

"No, you must come to the fort."

"We'd be in the way."

"A good many families are there now, and more are coming."

"But you say the French army has gone."

"Yes; it's gone. But not very far, and everybody knows it'll soon be back."

"We never could do it as things are now. The snow is deep, and is melting fast, and my wife and children could not travel on foot."

"Yes, ye can do it. I don't say 's how ye've got to go straight back with us, but ye've jest got t' come soon. It'll never do in the world for ye t' stay out here. It's worse 'n temptin' Providence."

"What makes you say that?"

"'Cause it's true! After what's happened here in the last few days, it's only common-sense t' know that's only a flea bite t' what's goin' t' be. I tell ye that this summer's goin' t' see the worst goin's on 'round here that's ever been."

"I think you are right," said the settler, soberly.

"If I could take my family now, I'd start for the fort. I know it's just as you say it is."

"Why don't ye go back with us?"

"When?"

"In the mornin'."

"We couldn't do it."

"Yes, ye could do it."

"I don't see how."

"Got any sleds here?"

"One bob."

"Flat runners or round?"

"Flat."

"Good. That's the thing. Put yer wife an' babies on that, an' we'll drag 'em all th' way t' Fort William Henry. When ye get there ye can rest easy, for if sixteen hundred Frenchmen an' Indians couldn't take it when there was only three hundred an' fifty men inside, what 'll they do when it comes to havin' all th' men we're expectin' there? Will ye do it? We'll stay here for th' night, an' all set out in the mornin'."

Jeremiah's intense earnestness was not without its effect; and before the men retired for the night it was agreed that all should set out on the following morning for the fort, whither many of the scattered families had already fled for safety.

Then Peter found his opportunity, and seated beside Sarah in one corner of the room he began to question her eagerly concerning the purpose of the hunter in departing from the place, and to explain also the plan proposed by Jeremiah.

"I don't think Sam would approve," said Sarah, thoughtfully. "He told us to stay here till he came back."

"When was he expecting to come?"

"He said it might be one day and it might be two, but he'd be certain to come."

"He may be here to-morrow morning before we start," suggested Peter. "And besides, I don't believe he'd oppose your going if he knew how many men there were here to help. You see, there are seven of us besides your uncle and Timothy Buffum."

"Timothy Buffum!" snapped the girl. "I'd rather have my aunt or the baby! He's the biggest baby of all. He's so afraid something 'll happen to him, he didn't even dare go with Sam. We'll have to look after him more than after my aunt!"

"Hush, he'll hear you."

"I don't care if he does! I tell you, Peter, I've had my hands full the past few days, and of all I've had to stand, he's the worst."

"Did Sam say where he was going?" said Peter, desirous of diverting Sarah's mind from the man for whom he himself cherished a feeling of contempt.

"He didn't say exactly, but I think you know as well as I where he'd be likely to go."

" The fort ? "

" Of course."

" Strange he didn't see anything of us."

" Probably he went by another way."

" Probably he did. What did he say because I didn't come back ? "

" Nothing much."

" Did he blame me ? "

" Blame you? How could he do that? You were doing everything in your power for us. I'd like to hear him say such a word! But he couldn't," she added quickly.

Peter laughed quietly, for it pleased him greatly to hear himself defended in this manner. " Did Sam leave any word for me ? " he inquired.

" Not a word."

" That's strange."

" Oh, you're not half so important as you think you are. We got along very well after you left us."

Peter's heart sank again. He could not understand how Sarah could talk in this manner, first taking his part sharply against any implied fault-finding by the hunter, and then quickly giving him to understand that his aid was by no means necessary to the comfort or well-being of the party. He was still puzzling himself over the matter when with the rest of the men he sought his quarters in the loft and the lonely little cabin was wrapt in silence.

On the following morning the rain began to fall, and the storm increased as the day passed. To return to the fort then was not to be thought of, and the men busied themselves in preparing the sled and such various belongings as they meant to try to take with them. For some reason, which Peter was unable to understand, Sarah seemed to take special pleasure in the company of John Rogerson, and to avoid him. This made him moody throughout the day, and he heartily wished that John had not been included in the number of those selected to accompany him on his return to the cabin. But somehow the long hours at last passed, and on the following morning the sun was shining brightly, and it was decided to set out at once for the fort.

The rude sled was loaded with the few articles of clothing the family were to take, and then the woman and the three children were placed carefully upon it. Sarah declared that she herself would walk, and as it was well-nigh impossible to find a seat for her on the sled, her decision was applauded by all. Making a virtue of necessity was not unknown, even in the days of the colonies.

Long, heavy straps had been fastened to the pole of the sled, and three of the men were to assist in dragging by pulling upon these while four were to pull on the pole itself. One was chosen to go in advance of the party to be on the lookout for danger and to select the easiest pathway, while another was to follow as a rear-guard and give warning of any danger threatening from that direction. Jeremiah's first plan had been to have a man also proceed on either side of the party, but the necessity of having all the assistance possible in hauling the heavy load made this impracticable.

To Peter first was assigned the duty of going in advance, and with his gun in his hands ready for instant use he entered upon his task. Jeremiah himself was the rear-guard, and for an hour or more the way was steadily maintained. It was slow and difficult work to drag the heavily laden sled through the snow, which was wet and soft from the recent rain; and the men were rejoiced when a halt was called and an opportunity afforded for a brief rest.

When the journey was resumed, Jeremiah took his place beside the pole and John Rogerson was sent to the rear, but Peter was still to go in advance, as his greater knowledge of the immediate region furnished ample reason for leaving it to him to select the way. Frequently he turned back to make a suggestion concerning the direction to be followed, then resumed his lonely march, constantly on the alert for any sight or sound that might indicate the approach or presence of the foe.

At noon time, when a longer halt was made, it was estimated that half the required journey had been accomplished, and thus far not a sign of the presence of enemies had been discovered. Though the men were wearied by the heavy task of the morning, their spirits were high, and it was confidently believed that before nightfall all would be in safety behind the defences of Fort William Henry.

After the hour of rest had passed, preparations were made for resuming the journey. The heavy sled was loaded once more, and the woman and children resumed their places. They were on the brow of a hill, and before them extended a long slope, where the task of dragging the sled would be comparatively light. Timothy Buffum advanced to the pole, and lifting it, pulled slightly upon it. The force was sufficient to start the unwieldy sled on its way, and in a moment it had gathered strong headway, and was swiftly passing down the hillside. Timothy's effort to stop it had been vain, and in a moment he was flung aside as it darted forward. A call to the men, who were near, but unaware of Timothy's action, instantly summoned them, and they began to run in pursuit of the fleeing sled, the motion of which was increasing rapidly. There was a shrill cry from the children, an exclamation of anger from the men, and then a sudden excitement which swallowed up every other thought.

CHAPTER XIX

An Ambush

AT the foot of the hill was a thick clump of young trees and bushes, and as the huge sled swerved slightly from its direct course down the hillside, it turned into the tangled brush. It was instantly overturned, and the occupants and load alike were flung far out into the snow. The mishap was serious, and the men ran to aid their friends; but just as they had almost reached the clump of trees a shrill cry broke in upon the silence. Instantly every man halted and drew his gun to his shoulder.

From within the hiding-place which the cedars provided darted several painted savages, whose sudden and unexpected appearance threw the men into confusion. The wild cries were answered by shouts from farther out in the forest, and for a moment it seemed as if the entire party had been surrounded by a band that perhaps had been following it since the time when it had set forth from the cabin.

The men darted for the shelter of the near-by trees, every one eager to gain a place where he would be protected and at the same time would

be able to use his gun to advantage; for the Indian
method of defence was common among the settlers.
The reports of the guns rang out, the wild cries of
the savages were redoubled, and for a moment the
excitement was indescribable.   Peter Van de Bogert
had been in the front of those who had been following the sled in its swift descent, and was also one of
the first to gain the desired shelter; but when he
looked before him not an Indian could be seen, uor
was the yelling repeated.   Suddenly, and almost
as if the earth had opened and swallowed up the
savages, the former silence returned.   Even the
white men had disappeared, for all had succeeded
in gaining the shelter they had sought, and were
hidden behind the great trees of the forest.

Peter's heart was beating furiously and he was
trembling with excitement.   He waited, but the
passing seconds might have been hours as far as his
perception was concerned.   He longed to dart
ahead and discover what the fate of the woman
and children had been, but he dared not move from
his shelter.   The leafless trees were motionless, the
deep silence was itself terrifying, and it required
all his self-control to prevent himself from shouting.

Only a brief time elapsed before Jeremiah Stubbs,
abandoning all thoughts of his own safety, leaped
forth from behind the tree which concealed him and
called: " Follow me!   We'll get 'em yet!   Come
on ! "

The men promptly obeyed the call of their leader and ran to the place where the sled had been over-turned. The pole had been wrenched from its place and was broken, and all the articles that had been carefully packed on the sled were lying where they had been scattered on the snow, but there was not a sign of the presence of a living person.

Jeremiah had been studying the footprints in the snow, and in a moment said, "They've taken the woman and the children, an' have got the man, too, and have put off."

"Well, we can follow them!" exclaimed Peter, eagerly.

"We can, but it's doubtful if we ever come back if we try it."

"How many are there?"

"Six or eight."

"We can get them, then! Let's try it! Come on! What are we waiting for?" demanded Peter, who was almost beside himself.

"Ye heard the yell out in the woods, didn't ye?"

"Yes."

"Well, that means there's a good many more there. They've got the whole lot prisoners, an' if we don't look out, they'll have us too. My 'pinion is we'd better put straight for the fort. We can't help what's been done hereabouts, but we can keep 'em from gettin' us too."

"And what good 'll that do?" demanded Sarah's

uncle, whose face was colorless, and whose hands trembled so that he could scarcely hold his gun.

"It's too bad," said Jeremiah, shaking his head sadly, "but it can't be helped. I'd be the first man to go 'f I thought 'twould be any use. Ye can rest easy that they won't harm 'em. They'll be treated 'most 's well as if they were in Fort William Henry itself. General Webb 'll 'tend to it for ye later, an' I make no doubt ye'll have 'em all back again safe an' sound."

"Never!" exclaimed the man. "I'll follow 'em if I have to go clear to Montreal."

"I don't blame ye fer yer feelin's a mite," said Jeremiah, sympathetically; "but it won't do a speck o' good. Ye'd better go back with us an' try what can be done from the fort."

"I'm going! I'll not stand here chattering like a lot o' crows."

The helplessness of the man, his grief and rage, were pathetic, and there was not one of the little band unmoved by his appearance. Sarah herself had now joined the men, but she had not spoken.

"Let him go, Jeremiah," said Peter, "and let me go with him. We can try to find out if they've gone to the Frenchmen's fort, anyway. The rest of you can go back if you want to, but I'll go on and help."

"Go on then, an' the good Lord go with ye!" said Jeremiah, solemnly.

Peter and his companion darted into the forest in the direction in which the prisoners must have been taken. The footprints could be easily seen in the soft snow, but Peter noticed that there were marks of only one of the children. He hastily concluded that the younger ones were being carried on the shoulders of some of the captors. The thought brought a measure of comfort, at least, that the design of the savages was to take their captives with them and not, as he at first had feared, treat the helpless ones as many before this time had been served.

The two men ran along the plainly marked way which the Indians had made no attempt to conceal. This fact was not comforting, for it indicated confidence on the part of the red men — a confidence which Peter was convinced could be born only of numbers. As he advanced, he glanced about him, fearful of sights and sounds that would bode little good. He could understand his companion's eagerness, and shared in his spirit, but he was in no wise deceived as to the peril in which both were. His one great fear was not of what lay before them, but of what might befall them from behind; for doubtless the Indians would expect to be followed, and would divide and arrange for a part of their number to fall upon any pursuing party from the rear.

For a moment both men stopped as a mark was

discovered in the snow which indicated that the woman had fallen.

"My poor wife! My poor wife!" moaned the man.

"She is all right. She's gone on, you see. There's no mark of trouble here," suggested Peter, who knew as well as his companion how the Indians were accustomed to treat those of their prisoners who were unable to maintain the pace at which their captors were moving.

"Come on! We must make haste! Come — "

The man stopped abruptly and pitched forward into the snow as the report of guns suddenly rang out from behind them. The expected had happened, and one glance at his fallen companion was sufficient to show Peter that the pursuit was at an end.

He turned and ran swiftly into the forest. Leaping over the fallen trees, plunging through the wet and heavy snow, disregarding the marks he was leaving, he ran on and on, mindful only of the one wild purpose of escaping from those who were, as he feared, close behind. He did not even pause to discover if he was being pursued, but fled for his life. Twice he slipped and fell headlong in the snow, and as he struggled to his feet it was with the conviction that some one had seized him. Finding himself still free, he resumed his breathless flight. Miles lay between him and the place where

his companion had fallen.   The sun had disappeared
from sight, and still the wearied and well-nigh exhausted youth fled from his unseen enemies.   Never
before had the terror of the great silence of the forest impressed him as in this flight.   The moon rose
full and clear, making it almost as light as day about
him, but still he fled.   Breathless at last, he stopped
and gazed almost stupidly before him.   He had, in
his flight, come to the border of the little clearing,
and at first did not recognize it as the place from
which the party had set forth that very morning.

In a brief time, however, he knew where he was,
and running among the charred stumps stood before
the door in the rear of the house.   It was fast, and
would not yield to the desperate attempts he made
to open it.   One of the last duties of Sarah had
been to secure the door when they had all gone out
from the building, and her work had been well
done.

The fear of pursuit was still upon him, although
he had seen no signs of the presence of his enemies.
With eager haste he darted around the corner of
the cabin and tried the door on the opposite side,
but that too was fast, and he could not secure an
entrance.   He glanced at the windows, but they
were too small to permit him to enter through them.
Something must be done, and at once; and almost
beside himself, he looked keenly all about him, hoping to find some means of assistance.   Suddenly his

eye fell upon two hop-poles near the well. He
looked at the low roof, and exclaimed: "I'll do it!
That's the very thing!" He placed the longer pole
against the eaves, scrambled up to the roof, and
drew the pole up after him. His gun had been
tossed up on the roof before he had made the
ascent, and now he ran with it to the huge stone
chimney and tried to look down into the darkness
beneath him.

It was but the work of a moment to drop the
hop-pole, lengthwise, into the open chimney. He
could hear it as its end struck the stone floor
below, and with his gun in one hand he seized the
pole and slid swiftly into the open fireplace of the
room beneath. His next task was to pull the pole
after him, but it was too long, and he was not able
to do it. It would not do to leave it in its place,
and he began to look about him in the dim light
for an axe. To his delight he found one, and then
with a few sharp blows he cut several feet from
the pole. The operation was repeated, and soon
the pieces of the hop-pole were on the floor beside
him, and he was relieved of one fear.

He next went to the small windows, but nothing
save the grim outlines of stumps and trees could be
seen. He listened, but not a sound lightened the
oppressive silence. He was alone, or if his enemies
had pursued him to the clearing, they were nowhere
to be seen.

Not yet convinced that he was safe, he carefully looked to the doors of the house. These were securely fastened, as he had already discovered before he had entered, and at last he seated himself with his back against the wall, prepared to wait and watch till the night passed. How familiar every object in the room was, and yet how changed! The rude bed where the sick woman had lain was still in the corner, the few wooden chairs were just where they had been placed by Sarah, but what a difference there was now that he was the sole occupant of the house! For a time he tried to conjecture what had befallen the captives. He trembled as he thought of the possible fate of the poor sick woman and the helpless little children. And Timothy Buffum was a prisoner, too. Doubtless he would never have a care for his unfortunate companions in misery, Peter thought, somewhat bitterly, for the man's intense selfishness was fully known.

He was still thinking of these things when at last his head began to nod, and despite his anxiety over his present peril, worn out by the exertions of the day, he was soon sleeping soundly. It was broad daylight when his eyes opened and he realized where he was and how he came to be there. He rose and crossed to the window to look for any signs of his enemies. The yellow sunlight flooded everything with its brightness, but to all appearances there was no one in or near the clearing but himself.

The thought was somewhat reassuring, and Peter began to search the house for something to satisfy his hunger. He discovered a piece of bacon and soon had a fire over which he was cooking the tempting food, his thoughts somewhat divided between his present occupation and his plans for the day, for soon he must decide what he would do. To attempt to follow the party with the prisoners was not to be thought of, and to return to the place where Sarah's uncle had fallen was equally out of the question. Still, something must be done, and soon. He turned the sputtering bacon, and was about to draw it from the fire when his ears were saluted by a sound that instantly banished all thoughts and questions from his mind. Something or some one was at the door in the rear of the cabin, and seizing his gun Peter quickly moved across the room.

CHAPTER XX

A GLIMPSE AT THE ARMIES

IN order that the events of the year may be more clearly understood, it will be necessary for us to leave the thread of this story for the moment, and consider some of the forces and elements that were working together to help the French commander, Montcalm, in his efforts to gain control of the fort on Lake George, which was held by the English colonists, and at the same time to open up the way which led to Albany and the regions beyond. As we know, the waterway by Lake George and Lake Champlain was at this time looked upon as essential to both sides. It was the easiest and most natural route into Canada, and was not wholly under the control of either the French or the English.

Young General Montcalm, clear-headed, earnest, light-hearted, and brave, had found soon after his arrival at Montreal that however popular he might be among his own men, the governor of Canada was not disposed to look upon him with favor. Indeed, the governor was sensitive because he himself had not been left in the chief command, and in his speeches and letters alike was inclined to assume all credit for whatever success was gained and to belittle the ability of the new commander.

However, Montcalm almost laughingly ignored the condescension of the governor, and began to carry out his own plans for the coming campaign. Of his own troops he had no fear; but Canadian forces were also required, and every effort was made to secure as large a force of Indians as possible to enter into the struggle. Quietly, persistently, the young commander labored, all the time retaining an appearance of light-heartedness, and choosing to ignore the slights and jealousy of his comrade-in-arms, who, despite his sensitiveness, was also doing his utmost to secure and equip a large force, so that the blow which it was planned to strike might prove to be a decisive one.

Throughout the winter the emissaries of the French were busily engaged in striving to persuade the Indians to join them in their expedition. Among the mission Indians (so called because of the mission work of the Jesuits amongst them — a work which had been largely successful, and one that had tied the red men to them by the strongest of bonds) the response was prompt and cordial; but it was harder to secure the aid of the more distant tribes, who naturally were more suspicious and less under the direct influence of the Jesuits.

However, patience, tact, promises of limitless re-
wards, the assurance of certain success, and the
action of neighboring tribes, at last succeeded in
winning the consent of the red men; and as the
summer days drew near, Montcalm found great
numbers of his savage allies gathering at Montreal.
Doubtless there were many misgivings in the heart
of the young commander as he watched the assembling of the hordes, for the warriors were as
impulsive as children, untrained to obey, and every
tribe sufficient unto itself.   Montcalm himself, un-
familiar with the customs of the red men, was
probably persuaded that all would be well; and
the supreme purpose for which he had come — the
capture of the English strongholds on Lake George
and in the near-by region — was enough to over-
balance any fears that may have arisen in his
mind.

As the warmer days drew near two facts were
learned by the vigilant French leader, which caused
his hopes to become high.   One was that the
English colonists had but very few Indians to aid
them.   This was due in part to the past action of
the settlers which had never been of a nature to
win the friendship, or even the good-will, of the
red men.   Then, too, the very confidence which the
English had in their own overwhelming numbers,
and their ignoring of all other things save the securing of lands, had led to a neglect of the ordinary

means of defence; and that, combined with the recent successes of the Frenchmen, had led the Indians to believe that the English settlers were inferior to their white neighbors of the north. The red men were probably as prone to be governed by the precept that "nothing succeeds like success" as their more highly civilized neighbors from across the great sea had been.

The chief source of confidence in Montcalm's mind, however, arose from the discovery that a large part of the English army had been withdrawn from the interior of the colonies and was to be sent against Louisburg. There was every prospect that they would succeed there, but meanwhile the waterway by Lake George would be poorly defended, and against the weak force that was trying to hold it he would send his stronger army. Already he dreamed of the success which he soon would win.

Montcalm had seized the little fort at Oswego, and the success there had had much to do in inducing the red men to side with him. One of the great chiefs from the upper lake region, after arriving with his warriors at Montreal, had said in a speech to Montcalm: "We wanted to see this famous man who tramples the English under his feet. We thought we should find him so tall that his head would be lost in the clouds. But you are a little man, my Father. It is when we look into your eyes that we see the greatness of the pine tree and the fire of the eagle."

Montcalm, ready with his warm words of welcome, was among those who moved about in the camps of the red men as they assembled, and his coming was hailed with the singing of war-songs and the feasts for which he himself had provided. The nearly naked savages, with their tall, straight, powerful bodies, their heads shaved except for the scalp-lock which was adorned with feathers, the huge brass earrings that had in some cases stretched the ear until the lobe came down almost to the shoulder, the beaver-skin blankets, the spears and bows and arrows with their quivers made of skin, —all these things must have made a strong impression upon the young commander, to whom they were new and novel.

He speedily discovered that difficult as it had been to secure the presence of these savage warriors, it was still more difficult to retain them after they had joined his forces.

They were like children in their eagerness for presents, and whatever they saw in the camp that pleased their fancy they immediately demanded. To refuse was to make them sulky and morose, and to give them what they desired, was to impoverish all. Even their food, which Montcalm found it exceedingly and increasingly difficult to provide, was a source of continual trouble. After he had care-

fully estimated and then given out the rations
which were considered sufficient for a week, the
red men would live in gluttony for two or three
days, and then come demanding that more food
should be given them.

Their superstition was also a source of anxiety.
If some chief, suffering from the effect of over-
eating, should have a "bad dream," he would at-
tribute his feeling to the work of evil spirits, and
his ever ready followers would declare that the
hand of the Manitou was against the proposed
expedition. Indeed, Montcalm felt that they vir-
tually were the masters and he the servant or even
the slave, for there was no escape from their per-
petual and unreasonable demands.

Of the eight thousand fighting men under Montcalm's command, nearly two thousand were Indians,
gathered from the far west as well as from the
more immediate region, speaking different languages,
with differing customs, and apparently alike only
in their ferocity and childish demands.

"War feasts" had to be carefully observed, or
trouble was likely to ensue. The painted savages,
seated in two lines, each facing the other, would
be served or serve themselves from the huge kettles
filled with meat that had been placed in the space
between the two. For a time a dignified and
impressive silence would be observed, and then
would follow a wild song, the burden of which

was, "Let us trample the English under our feet,"
repeated again and again.   Next some warrior
chieftain, filled with a boastful spirit, would rise
and begin to recount vividly the brave deeds of his
tribe.   At these feasts it was the custom to name
the chiefs who were to lead the red men into the
battle.   As each was named he rose, stepped forth
from his place in the lines, and holding aloft in his
hand the head of some slaughtered animal, shouted,
"Behold the head of the enemy!"   Loud cries of
approval from the braves followed, and then the
chief who had been named marched back and forth
singing his own praises and boasting of what he
and his followers soon would do to the English.
Deep, guttural ejaculations of approval rose from
the warriors as he passed; and as no limit was
placed to the time in which each chief might display himself and boast of his mighty deeds, it was
only natural that these feasts should be prolonged
and repeated until every one had had his opportunity to impress all the beholders with the full measure of his own importance.   Guns were provided
for these warriors, but they became only fairly
skilful in the use of them.

    With the approach of summer days, fleets of
canoes or bateaux began to move up the uncertain
waters of Lake Champlain, and the destination of
all, white men or red, was the fort at Ticonderoga.
Not merely was the little fort itself filled with men,

but forces were stationed at various places in the immediate vicinity. And the white men were toiling with tireless energy. Improvised sawmills were busy, food for a period of a month was being collected for the great army, and the restless red men, though they took no part in the manual labor, were employed in scouting, or sent to capture some party of stragglers that had wandered too far from Fort William Henry. Indeed, Englishmen and Frenchmen were alike dependent upon scouts or prisoners for information as to the plans of their enemies, and each was continually sending parties into the region occupied by the other for the purpose of securing those who could be compelled to relate all they knew.

Meanwhile, at Fort William Henry the men were not idle. The fort was made in the form of a rude square. Heavy logs, laid in tiers, with the spaces between filled with earth, were placed above the embankments. On one side was a great swamp, on another was the lake, and ditches had been dug on the other two sides. Seventeen cannon, including some that were quite small, were mounted on the walls, but the strongest defence the little fort knew was in the sturdy Scotchman who was in command, Lieutenant-Colonel Monro.

Fourteen miles from Fort William Henry was Fort Edward, where General Webb, with a force a trifle larger than that behind the defences of the

little point on the lake, was in command.   Upon
him really rested the defence of both forts, and
when it became known that the French force,
greatly outnumbering that of the provincials, was
near at hand, he wrote the governor, begging that
men might be sent to his aid.   But many of the
regulars had gone with Loudon to Louisburg, and
of the militia there were not many to respond.
Besides, it was not yet evident that the French
were so deadly in earnest and that the peril was as
great as the men in the two forts believed.

The condition of the English troops was not such
as to warrant great confidence.   Unaccustomed to
the demands of military life, many of them were
ignorant of the ordinary means of retaining their
health in such places.   Sickness was prevalent, and
even smallpox came to add to their miseries.   If
the word of some of the chaplains is to be received,
it is to be feared that the moral status was not a
vast improvement over the physical.   "Curseing
and swareing," one of them complains, were prevalent.   There was a disregard of the "means of
grace," and, all together, the poor chaplains seem to
have had a hard time of it.

And yet the men in Fort William Henry were
sturdy and true, neither better nor worse than their
fellows.   They realized that upon their efforts rested
the defence of the region from the advancing Frenchmen and Indians.   That Loudon had withdrawn so

many of the men to go to Louisburg did not detract
from the seriousness of the condition in either of
the two forts, and as the summer drew nigh, every
one was aware that grave problems were to be
solved, though as yet no one believed them so
weighty as they were soon proved to be.   General
Webb was still begging for help, Colonel Monro
was doing his utmost to strengthen the fort under
his command, and both leaders were striving to
provide for the wants of the straggling families
that almost daily were flocking to the forts for
shelter and protection.

CHAPTER XXI

In the Cabin

IT was long before the events recorded in the preceding chapter had been brought to pass, however, when Peter Van de Bogert, startled by the unexpected sound at the door of the hut, advanced and stood ready to open it.

"Who is it?" he called in a low voice.

"It's some one you know, Peter," came the response.

"Are you all right and all alone, Sam?" inquired Peter, for he had already suspected who his early visitor was, and was therefore not surprised when he recognized the voice of his friend.

"All right and all alone," responded Sam.

As he spoke, Peter had lifted the heavy bar from its place and the door was swung open. "I tell you, Sam, I never was so glad to see any one in my life!" he said, as the hunter entered and the door was once more barred. "We've had some terrible times since you went away."

"So I hear, so I hear."

"Have you seen any one? Did you know that

214

Sarah's aunt and all the children and Tim Buffum
are all in the hands of the redskins?"

"So I heerd."

"Then you've seen some of them, have you?
Have you been to the fort? How is everything
there? Any more signs of the Frenchmen?"

"The thing for us to do now is first to put ourselves outside that bacon ye're cookin'. I don't
mind tellin' ye I've got somethin' t' say to ye, but
ye better get yer breakfast first."

"Is it bad news, Sam?" inquired Peter,
anxiously.

"That's as may be. Some might call it one
an' some t'other."

"Anything about the prisoners?"

"Not exac'ly."

"Tell me what it is, Sam."

"Jest 's soon 's I've 'tended t' this bacon. Got
any fixin's t' go with it?"

"I can't find anything but the bacon."

"It's goin' t' rain," remarked Sam, as he busied
himself in preparing their breakfast, a task that he
undertook at once. "There, jest hear that, will
ye!" he added, as there came a heavy shower.
Though impatient to receive the information hinted
at, Peter understood his friend so well that he
knew it was useless to attempt to obtain it from
him until he was ready to give it of his own free
will. Accordingly he made no response to Sam's

declaration, though he could see as he glanced through the little window that the ruins of the barn were almost hidden from sight by the downpour.

"Looks little 's if we'd be here for quite a long spell, Peter," remarked Sam, as at last, apparently satisfied with his observations, he drew the smoking, sputtering bacon from its place over the fire.

"I hope not," responded Peter, gloomily. "I've seen all I want of it."

"Sort o' miss some o' th' people that b'long here, so t' speak?"

"I didn't say that."

"Didn't need to."

"Sam, how did you stave off the Indians from getting in here? Did you wonder why I didn't come back to help?"

"I didn't stave 'em off, we just kept 'em out. I kept my eye on 'em day an' night, but when the Frenchmen an' redskins quit tryin' t' break into Fort William Henry and cleared out, all those out here quit 'long with 'em. I've never yet seen th' thing I couldn't handle if I only had the patience. I tell ye, Peter, patience 's the biggest little thing in the world. How'd I ever get a mink 'f I wasn't patient? Why jest look at a catamount! He's the most patientest thing there is in the whole woods. He'll wait an' wait *an'* wait till it's jest the right

time for him to strike, an' then! Why when he
does strike, it's jest 's if he never had a bit o'
patience in all his life. Now it's the same way
with people —"

"Yes, I know what you think about that, Sam.
I've no doubt you're right; but what I want to
know now is, how it is that you are here, and
where you've been, and whether or not you've
heard anything about the prisoners."

"Ye remember what I said 'bout patience, don't
ye, Peter?"

"Yes."

"Well, don't ye forget it, then." There was a
quizzical expression in Sam's eyes, and Peter knew
he was about to hear of what he was most eager to
learn. But he did not speak.

"Well, when ye didn't come back, Peter, as
'twas agreed ye'd do, I knew one o' two things
was the matter; either ye'd got into trouble yerself, or else, if ye'd made yer way into the fort
all right, then there was trouble there an' they
wouldn't let anybody come back with ye. There
was jest one thing, though, that I was disap-
p'inted in."

"What's that?"

"That ye didn't come back yerself, when ye
found out they wouldn't send anybody back
with ye."

"The major wouldn't let me come."

"Th' major wouldn't 'let' ye? Shucks! I'd
like t' see him stop me!"

"That's different."

"No, 'tisn't dif'rent. I tell ye, one man's jest
's good as another, an' I don't ask anybody if I can
go when I want ter, or come when I like."

"Suppose they all did that way, Sam," suggested
Peter, who was well aware of the prejudice and
also of the practice of the hunter. Accustomed as
Sam was to the life of absolute freedom in the
woods, he scornfully rejected all the discipline and
restraints of army life, though he was as eager as
the most fervid patriot in his desire to hold back
the encroaching Frenchmen. He would not regu-
larly enlist in the service, but he was so well known
by the leaders, and his knowledge of the woods and
his skill in meeting the problems of the border were
so great, that his services had been accepted almost
upon his own terms.

"Sam, was there any news from Sarah's aunt or
the children?"

"Not a word, except that they're gone."

"Do you think they'll ever get back?"

"That's as may be. It'll all depend, in my
'pinion, who 'tis that's got 'em, an' whether they're
taken to the fort or not."

"The fort? What fort?"

"What fort do ye suppose? Ye didn't think
they'd bring 'em up t' Fort William Henry, did

ye ? No; there's only one fort where they'd go, an' my only hope is that they've been taken there."

"You don't think they have been, Sam. I can see it in your face."

"I don't know what t' think, lad, an' that's a fact. They're gone, an' that's 'bout all we know 'bout it. We'll jest have t' hope for th' best, an' rest on that." Sam was speaking soberly, and Peter could see that he was in no wise hopeful as to the outlook for the unfortunate prisoners.

"Sam," he said abruptly, "how is it that you are here now? You haven't explained yet, and I've been patient, too."

"So ye have, lad, so ye have. I can't deny that."

"Well, why is it, Sam ? "

"D'ye hear that rain ? " demanded the hunter, abruptly, as the downpour suddenly increased. "That 'll help our plans a bit, an' yet 'twill hinder 'em, too. I'm thinkin' we'll have t' have a little more patience yet."

"Mine is almost gone."

"I'll lend ye some o' mine."

"Keep it! I don't want it!" retorted Peter, sharply. "I'm going to start for the fort pretty soon."

"Better wait till it slacks up a bit. Might get wet if ye started now. I guess we'll have time 'nough for me t' explain things, seein' as how we're likely t' be kept here for two or three days."

"Two or three days? I don't understand what you mean, Sam."

"I came here 'cause I knew you was here."

"Yes?"

"I was on my way from the fort, an' met th' party that you was with an' that started from here. An' then I went back with 'em t' the fort."

"You did?" exclaimed Peter, instantly interested.

"That's what I did. But I think it's likely t' be some time before either you or I'll see Fort William Henry again."

"What do you mean?"

"Jest what I say. Don't ye think I'm in th' habit o' that?"

"Yes, yes. But I —"

"Ye will, then, pretty soon," interrupted Sam, laughing lightly. "The plain state o' the case is that Major Eyre says Colonel Monro wants us t' go down th' lakes — Whew! Hear that rain! It 'll be good-by to all the snow an' ice if this thing keeps on!"

"Go on with your story, Sam," said Peter, impatiently.

"As I was sayin' when ye interrupted me, we're to go down th' lakes."

"Both lakes?"

"There's only two, 'cordin' t' my way o' thinkin'."

"Do you mean that we are to go clear to Canada?"

"That's as may be. What we're to do is to
keep on till we find out somethin' worth findin'
out, an' then we're to put straight back to William
Henry with the news."

"Do you think we'll get past the Frenchmen?"

"'Past' 'em? No. We don't want to get 'past'
'em, leastwise past 'em all. What the colonel
wants, 's I understand, is for us to go just as far
as we can go down Champlain, an' find out all we
can about what's goin' on there an' what's likely t'
be done. He knows about Ti, but what he doesn't
know is how many are comin' an' what th' next
plan is likely t' be. He's got a good many men
out hereabouts, an' he's goin' t' send some more
besides us six clear to the end o' Champlain — "

"Six? Did you say there were to be six?"

"That's it. Ye guessed right the very first time
That's exac'ly what I said."

"Who are they?"

"There's you an' me — that makes two. Then
there's t' be Jeremiah Stubbs an' that young John
Rogerson. I guess I ought t' have said five an' a
half, for I don't b'lieve that young chap 'll count for
more'n a half a man, leastwise unless he's picked
up a lot lately."

"You needn't be afraid of him. John Rogerson
will do his part every time, you can rest easy about
that," said Peter, warmly. "But who are the
other two? You've mentioned only four."

"Two o' th' redskins are goin' 'long.   There's Tehasen an' Totaro.   They'll both go."

"When are we to start?"

"That's as may be.   Depends a bit upon wind an' weather.   If the thaw had held off, we'd have started to-day, and tried it over the ice.   Seein' as how the thaw hasn't kept off, we may have to wait till the ice breaks up an' try our luck with canoes."

"Do you know just where you're going, Sam?" inquired Peter, anxiously.

"I know where I'm goin' t' try t' go."

"Where?"

"Just where I told ye — as far down Champlain as we can go.   Ye know, Peter, most o' the Frenchmen and redskins are comin' up the St. Lawrence and then up the lake, so we'll know 'bout where t' look out for 'em."

"Do you think we can ever do it?"

"We can try it."

"So we can, Sam.   I'm ready to do my best. Are we to go back to the fort to meet the others and start from there?"

"If this rain holds on we may have to, though we weren't exac'ly plannin' t' do that.   We'll have to wait an' see.   Sort o' give our patiences a chance to get into good workin' order first."

There was ample opportunity for that very excellent quality to have its perfect work before the raging storm subsided. For two days the rain fell in torrents, and the snow was almost entirely washed away before it ceased. The bare earth appeared, first in discolored patches and then in more extended places, until at last it was visible on every side and only occasional banks of snow remained. The very trees looked wretched and forlorn. Winter had passed, but the spring had not yet appeared.

On the morning of the third day the sun rose clear, the air was warm, and the disconsolate region was covered with brightness.

As the two inmates of the little cabin surveyed the scene before them, the hunter said: "It's settled for us, Peter, we'll go by canoe, for the ice must be out now. It never could stand such rain as we've had, and this sun. An' I'm glad o' it, too."

"So am I," responded Peter. "What are we to do now?"

"Nothin', till noon. Give our patiences a little more chance."

It was about ten o'clock, when both men were startled by a song which rose from the border of the forest.

> "Lucy Locket lost her pocket,
> Kitty Fisher found it.
> Ne'er a penny was there in it,
> 'Cept the binding round it."

CHAPTER XXII

THE DEPARTURE IN THE NIGHT

"THE fool!" muttered the hunter, as the sound of the song came clearly from amongst the trees. "He must be clean daft, whoever he is."

Peter was as anxious as his companion, although it was evident that there was nothing to fear from the newcomer, whoever he might prove to be, for no enemy would announce his arrival in such stentorian tones. Both men watched the border of the clearing for the man to appear, and soon John Rogerson stepped forth and boldly approached the cabin.

"Just 's I thought!" exclaimed Sam, in anger. "Wonder what the little sprint thinks he's tryin' t' do? Let all creation know what he's about?"

Apparently unmindful of the reception he might receive, John Rogerson moved swiftly toward the door.

"I've a good mind to lift his hat off his head jest t' show him what a foolish thing it is he's doin'," muttered Sam, as his finger rested upon the trigger of his gun. "How does he know this house isn't full o' Frenchmen?"

John, however, was now near, and again began
to sing : —

> "Lucy Locket lost her pocket,
> Kitty Fisher found it —"

"Stop! Stop right where ye be!" shouted Sam,
unable to restrain himself longer. "Jest give us
a word as to who ye might be."

"You know who I am all right," laughed John,
recognizing the voice of the trapper, and continuing
to advance. "I'm with you once again! Glad to
see you both," he added, as he approached the
door, which now had been flung wide open, showing Peter and Sam standing in the doorway.

"It's a good thing somebody's glad," said Sam,
glumly.

"Why, Sam, aren't you glad to have company?"
demanded John, laughingly. "I thought you'd be
homesick and lonesome, so I came all the way
from the fort just to keep you company. I'll go
back if you say the word."

John Rogerson's appearance mollified the hunter's
anger somewhat. He was bespattered with mud,
and it was evident that his journey had not by any
means been an easy or pleasant one.

"What d'ye make such a racket for, I'd like t'
know?" demanded Sam, sharply.

"I wanted to let you know that friends were
coming to your relief."

"Humph! Mighty little relief, I'm thinkin'. How d'ye know where we was?"

"I didn't know, but I suspected that you might be here. So I came to see."

"Well, all I can say is that the next time ye come, don't try to tell all creation that ye're on the way. Just s'pose the woods had been full o' redskins, same 's they was a few days ago."

"But they aren't."

"Humph! That's as may be. Ye always want t' be on yer guard and never give th' other side th' chance t' find out what ye're doin'. That's my plan pretty much always an' everywhere."

In spite of his woebegone appearance John's spirits were not to be checked, and he laughingly declared that he would not forget the hunter's words of warning; but he had been so well assured that the immediate danger was gone, that with the passing of the storm he had been unable to restrain his impulse to sing of Lucy Locket's misfortune.

In response to Sam's questions, John explained that although the ice in the lake was broken, it had not, as yet, all disappeared, and that he had come with word from Jeremiah Stubbs that it was deemed wiser for them all to assemble at Fort William Henry and delay the time of their departure for a few days, until a safe passage by water could be made.

"What's Jeremiah Stubbs got t' do with it?" demanded Sam, sharply.

"Not much; only he's to be one of the party, and as you didn't happen to be in the fort, perhaps the major or the colonel talked with him about the trip and made a suggestion or two. I don't know anything about it myself. I'm only tellin' you what was told me."

"That's all right. I guess Jeremiah 'n' I understand each other. D' he say for us t' come directly back t' th' fort?"

"He didn't say. He probably thought you knew best about that. They won't start till you get there, I'm sure o' that."

"That's as may be," assented Sam, his good nature now completely restored. He had not seen John's sly wink at Peter when he had been speaking, and his easily aroused jealousy over his own position as a leader in every plan was therefore quieted.

It was afternoon before the three men set forth on their return to the fort, and, graphic as John's description of his own journey had been, Peter and Sam speedily discovered that the conditions had not been overdrawn. The mud seemed to be well-nigh bottomless, and the pools of water, caused by the heavy rains and the melting snows, were in places impassable. Thoroughly wearied by their exertions, they at last arrived at the fort late in the afternoon in a condition similar to that in which John had been when he had entered the cabin.

A change of clothing and the rest of a night completely restored them, and on the following morning there was a conference of Sam and Jeremiah with Colonel Monro, while Peter and John sought out the place where Sarah was staying.

"Have you had any word from my aunt?" was Sarah's first and eager question when she had greeted her friends.

"Not a word. I was hoping that I'd hear good news from you," said Peter.

"No good news, no news at all," responded Sarah, sadly. "I should think that Colonel Monro would send a party out after them."

"I'm afraid it wouldn't do any good just now. Everybody seems to think they'll be taken care of, Sarah. You see, it isn't just as if the redskins were the only ones here. The Frenchmen are white men, and will see to it that no harm comes to them," said Peter.

"Oh, you'll hear good words from them pretty soon," said John, glibly. "A friend of mine lost something very valuable not long ago, but some one else found it for her, and she's been happy ever since."

"Who was it?" inquired Sarah.

"It was Lucy Locket. She lost her pocket. Perhaps you heard about it."

"Don't! Please don't, John!" said Sarah, her eyes filling with tears.

"Why, I didn't know — I didn't mean —" stammered John, taken back by the sight of Sarah's sorrow. "I guess Sam was right when he said I was a fool."

"I guess he was," assented Peter.

"You don't know anything about it, Peter Van de Bogert!" retorted John, angrily.

"No, I don't," said Peter, soberly. "I was just taking you at your own word."

"If I don't hear anything soon," said Sarah, quickly, "I think I shall go back home. My poor aunt! I don't know what I shall do without her!"

Peter did not speak of the rumors that were current in the fort, that the scattered people of the region were already fleeing for safety to Fort Edward or to Fort William Henry, and that the possibility of Sarah's return to Albany or to her own home was not promising. She would learn of all that in due time, and meanwhile her anxiety and sorrow over what had befallen her aunt and the children were so keen that all he could do was to try to console her by the general expression of his sympathy, and the prospect, which in his own heart he was not able to believe was very bright, that they would not be ill-treated in their captivity. That they were not the only ones who had been seized was slight comfort, for it is to be questioned if "misery does love company," in spite of the common declaration to that effect. But the conditions were unique, and there was no disguising the feeling of anxiety that pervaded all the inmates of the fort. There were harsh words for Loudon, who had withdrawn so much of the army and left the remnant to face an advancing and overwhelming force of Frenchmen and Indians.

It was three days afterward when Peter, with his five companions, three white men and the two Indians, prepared to depart from Fort William Henry on their perilous expedition. Their orders were somewhat general, and much was left to the discretion of the hunter and Jeremiah Stubbs; but the men all understood that they were to pass the French lines, go as far as was thought wise down Lake Champlain, obtain all the information they might be able to secure, and otherwise were to act as occasion demanded.

The ice was gone from the lake now, and the rays of the springtime sun were warm and golden. The nights, however, were still cool. It was decided that the first part of the attempt should be made in the darkness. If they should succeed in passing the French fort, then it might be wise to try to advance in the daytime, but until that had been accomplished, it was deemed best to paddle by night and conceal themselves by day. Three canoes were to be used, two men being assigned to each, and each canoe was to contain food for its occupants sufficient to last for several days, for it was not known how long the little party would be gone.

" I don' know 's I think much o' that plan," said Sam, when he learned that the two Indians were to go together in one canoe in advance, while he and Jeremiah followed in a second, and Peter and John were assigned to the third.

" What's th' trouble? Don't ye like yer comp'ny?" demanded Jeremiah.

" My comp'ny's all right enough," said Sam, " but I'm not over happy at leavin' those two boys together. First thing ye know, that there John will be yellin' out 'bout his ' Lucy Locket,' an' tellin' all Canada to look out, for we're comin'."

" John's jest 's good 's th' other boy," asserted Jeremiah, sturdily, for he had a strong affection for his light-hearted young comrade-in-arms.

" That's as may be," said Sam. " Time 'll tell."

" Let it tell, then. What we want t' do is t' get ready t' start."

The preparations were at last completed, and all the various articles to be taken with them were carefully stowed away in the light canoes. Their expedition had been kept secret, so that when at last the men were ready to depart, no one accompanied them to see the start. The serious nature of their undertaking, however, was sufficient to occupy their minds, especially those of the younger members of the party, and it was with a feeling of relief that they quietly left the fort and joined their comrades on the shore.

The darkness was not deep, and they could see far out over the waters of the lake before them. A few frogs could be heard, but otherwise there was nothing to break in upon the quiet of the night. Somehow the expedition appeared in a different light, now that they were about to start, and even John was silent.

The two Indians took their places in their canoe, and with long, noiseless sweeps darted out upon the water. Sam and Jeremiah followed them, and without a word Peter and John brought up the rear.

The three canoes, keeping well together, moved like shadows out upon the dark water, and, well in toward the shore, proceeded on their way. It had been decided that no great distance should be covered that night, but that the strength of the men should be carefully preserved, for no one knew at what moment all their skill and power would be demanded.

The low shore could be seen not far away, and the croaking of the frogs in the adjacent marsh provided the accompaniment to the rhythmical dip of the paddles. It seemed to Peter that not many miles had been covered, when he perceived Sam's canoe dropping back to them.

"What is it, Sam?" whispered Peter, when the canoes were alongside. "Anything wrong?"

"Jeremiah says we ought t' stop. Th' signs are all wrong."

"Where shall we stay?"

"I know a good place not far from here."

The conversation ceased, and the three canoes were headed for the shore. Before they arrived, Sam declared that he would land and learn if the place was suitable, and also if the signs were good, and, accordingly, the others waited for his return. In a brief time the hunter came back, reporting that all was favorable; and then all six, after they had carefully concealed their canoes, made their way to an abandoned log-house, the existence of which the hunter had known. A guard was set for the night, and on the following day Peter and John were left in the hut while their companions scattered to search for signs of the presence of their enemies.

Late in the afternoon all returned with the report that nothing of interest had been discovered, and soon after nightfall the voyage was resumed. For some reason the canoes were not kept so close together as on the night before, and several times Peter lost sight of his companions, although, in each case, he speedily overtook them.

Again the distance between them lengthened, but the former success had made the young soldiers somewhat indifferent, and they gave slight heed to the fact, until, as they were rounding a point, they beheld a canoe directly before them, and apparently waiting for them to draw nigh.

"What is it, Sam?" said Peter, in a low voice, as he came alongside.

Then without waiting for a reply, he suddenly fell back, as he perceived that the men in the other canoe were not the hunter and Jeremiah, but a white man and an Indian who had promptly seized the gunwale of his own canoe.

CHAPTER XXIII

Down the Lake

THE silence that ensued for an instant was so intense that Peter Van de Bogert, as he afterwards described it, "felt almost as if he could *hear* it." In the dim light he could see the faces of the men in the other canoe, and knew at once that they were not friends, and what their purpose was. A number of expedients to avoid capture flashed into Peter's mind, but not one seemed feasible. Not more than a hundred yards distant lay the shore. In advance were the two canoes bearing Sam and his comrade and the two Indians, but how far away they were it was impossible to know. To reach for his own gun was impossible, for the first movement of that kind on his part would cause the men beside him to make use of their weapons, which each was holding in his disengaged hand.

All these thoughts had flashed into Peter's mind in the brief instant while the two canoes were alongside and the men had been silent. It was almost like a wild dream. He was trying to escape, yet was fast held in the grip of his enemy.

Suddenly Peter realized that he must act at

once if anything was to be done. Help from his comrades might even now be impossible, and surely would be if he delayed longer. Without looking at John, Peter raised his voice, and, unmindful of the possible presence of danger on the near-by shore, gave one loud, long call : —

" Sa-a-m ! "

At the same time he leaped from his own canoe full against the man who was holding it. He made no attempt to seize his foe, for he was holding his own gun in his hands, having grasped it as he leaped. Instantly both canoes were overturned and all four men were thrown into the lake.

When Peter rose to the surface one of the canoes had righted itself and was floating near him. He instantly seized the bow and placed his gun, to which he was still clinging, inside. It was useless to attempt to crawl in, for the canoe was too light, and was easily upset, and even under the most favorable circumstances, it would have required more skill than he possessed to accomplish such a feat.

In the water near him was the head of a man, but whether it was that of friend or foe he could not determine. He swung the end of the canoe around, and propelling it by the motion of his feet soon had it where the other man could grasp it. As the face was lifted from the water he could see that it was that of the white man who had been with the Indian in the other canoe. For the moment, at least, Peter felt himself safe, and then he glanced hastily about for some indication of what had befallen his friend. That John was an expert swimmer he knew, and the knowledge had been one of the causes of his own action, for to his mind almost anything had seemed preferable to capture.

He soon discovered that John and the Indian had succeeded in gaining the other canoe and that each was holding to an end of it, just as he himself and the white man were doing. The situation certainly was unique, and under other circumstances Peter might even have seen its somewhat ludicrous aspect; but as it was, his teeth were chattering with the cold, he knew that the contents of each canoe had been lost in the overturn, and that doubtless the only gun in the possession of the party was his own, now lying useless in the bottom of the boat.

" Are you all right, John ? " he called.

" All right," responded his friend, " though I can't just say I'm happy."

The white man also called to the Indian, but as he spoke in French his words could not be understood. What he had said, however, speedily became apparent when, as if by a preconcerted plan, both men began to swing the canoes about, and by vigorous pushing endeavored to send them in the direction of the shore.

"Look out, John!" called Peter, sharply. "Do you see what they are up to? Don't let them take us ashore. Keep your end out in the lake!"

John evidently required no help, for at once he began to kick lustily; and as Peter also was busy in a similar task, the two canoes were soon making circles in the water as the young soldiers strove to hold them in their present positions. For a time the splashing and the noisy efforts of all four men continued, and not much headway was made toward the shore. Even the cold was forgotten in the violence of the exertions and the determination to avoid the capture which would surely follow if the Frenchman and the Indian should succeed in their attempts. After a time the struggle was abandoned, and for a few minutes the two canoes floated idly on the water, the man at each end cautiously watching his opponent, and prepared for action at the first sign of a new plan.

The situation was not one which could long be endured. Peter's hands were already becoming so numb with the cold that he knew he should soon be unable to retain his grasp. His sole consolation was that his adversary was doubtless as badly off as he, though the knowledge afforded no special comfort. If help did not come, he knew that he must let go and strive to make his way to the shore; but the moment he should attempt that he was confident that the man opposite would clamber

into the canoe and speedily overtake him, though
both paddles had been lost.   The gun, too, would
be a weapon which the enemy could use as a club,
and at the thought Peter clung more desperately to
his hold.

"All right yet, John?" he called, somewhat
loudly, for the distance between the two canoes
was greater now.

"Ye-e-s," came the response.   "What are we
going to d-o-o, Pe-eter?"

"Hang on, that's all we can do."

"But the other fellows are hanging on, too."

"Let them hang."

"I-I wi-i-sh the-ey w-w-ould!" chattered John.

The conversation ceased abruptly, as out from
the lake suddenly darted two canoes.   In an
instant they had come alongside the one to which
Peter was clinging, and as he looked up he
recognized the face of his friend the hunter.

"Here I am, Sam," he called.

"So I see.   Who's at the other end?"

"A Frenchman."

"Where's John?"

"With the other canoe."

Sam instantly turned and spoke a few words to
the two Indians who had followed in their canoe,
and together dipping their paddles into the water
they started toward the place where John could be
dimly seen.

"I don't dare to try to lift ye into th' canoe, lad," said Sam.

"D-do w-what you w-want t-to. I'm glad you came."

"Can ye hang on a bit longer?" inquired the hunter, anxiously.

"Y-y-es, I-I g-guess s-so."

"We'll take ye ashore, then, canoe an' all."

As he spoke the hunter swung the canoe about, and bidding Peter do his utmost to cling to his position, he at the same time tossed the end of a long strap to the white man; and then he and Jeremiah began to send their own canoe, with the other and its load in tow, straight for the shore. They speedily overtook the other two canoes, and, at a word from the hunter, the two friendly Indians followed his suggestion, and the little procession drew near the land, Sam and Jeremiah leading the way.

While the hunter assisted his friend to land, Jeremiah securely bound the hands of the white man, who could offer no resistance, both because he was so completely chilled by the icy waters and because any movement on his part would have been instantly seen by the keen and watchful ranger. The canoes were drawn up on the shore, and while Sam was attending to Peter, Jeremiah stood waiting to assist the men who were approaching in the other canoe. There had been a delay in its arrival, but in a brief time it approached and was drawn from the water by Jeremiah.

"Where's th' redskin?" demanded Jeremiah, sharply, as he saw no one besides John and the two Indians.

"Ugh!" replied one of the Indians, as he held up a fresh scalp.

For a moment even the ranger appeared shocked by the sight, but the hunter quickly said: "No use, Jeremiah. If ye're goin' to use redskins, then ye've jest got t' put up with redskins' ways. 'Tisn't your way nor mine, but I'm thinkin' it may be one less t' try the same trick on us, if what I hear about the number o' th' savages that are comin' 'long with this Montcalm is true."

"'Tisn't a good sign, though," muttered Jeremiah.

"That's as may be. It's a better sign, 'cordin' t' my way o' thinkin', than 'twould be t' see one o' our scalps a-hangin' from their belts. Say," he added, turning to the two Indians as he spoke, "you speak French. Jest ask this Frenchman if he happens to know if there's a camp anywhere hereabouts."

The red men reported that there was one a few miles farther down the shore, but none in the immediate vicinity.

"Then we'll have a fire," said the hunter.

"I'm agin that!" protested Jeremiah, earnestly.

"That's nothin' new, for ye're agin 'most everything, Jeremiah. I know what ye're thinkin' on, but 'twill be all right. We must give these men who've been in the water so long a chance to dry out or we'll be havin' somethin' worse 'n th' small-pox t' deal with. We shan't have t' stay here but a little while, an' 'f what this white man says is true, then if any one should happen t' smell th' smoke, he couldn't get here till long after we've gone on. Besides, both Tehasen an' Totaro will be on th' lookout for us. We've got t' take some chances in this world o' ours, an' that's jest what I'm goin' t' do now."

The hunter collected some broken branches of the fallen trees, and before long had a blazing fire under the shelter of a high bank. The prisoner and both young soldiers were soon enjoying its warmth, but after a half-hour had elapsed Sam declared that the place must be abandoned and the expedition resumed. " 'Twon't do t' stay more'n a minute longer," he said. " But we'll stop jest long 'nough t' hear Peter tell how it all come about. I didn't ask him before, for I wanted t' give him a chance t' warm up an' get rested."

Thus bidden, Peter briefly related the story of their adventure, and as he ended, Jeremiah said sharply: " There, Sam ! That's jest what I was tellin' ye. We ought not t' separate as we did. Those fellows were there on the lookout, an' jest waited till we'd got ahead o' th' boys, an' then they crept up, an' there ye was! "

" Yes, that's so, Jeremiah," assented the hunter. " Ye were right for once in yer life."

" Humph !" retorted the ranger. " Maybe 'nother time ye'll be ready t' listen."

" That's as may be."

" What ye goin' t' do with this man, Sam ?" inquired the ranger, glancing at the prisoner as he spoke.

" Better take him with us."

" Better send him back t' the fort 'long with Peter."

" Might send him back with John," said Sam, sharply.

" We'll have need o' John afore we're done, unless all signs fail."

" We'll need Peter more, I'm thinkin'."

" Sho ! John wouldn't 'a' tipped those canoes over an' lost the guns an' all. I tell ye what 'tis, Sam, we're goin' t' think o' that stuff we brought 'long t' eat a good many times before we're back at Fort William Henry."

" No, John wouldn't 'a' tipped the canoe over, but where'd he an' th' canoe, too, been by this time ? Just tell me that, will ye ? Ye know 's well 's I do, we'd 'a' lost not only the victuals, but canoes an' guns an' men, too !"

" That's all easy 'nough t' say now. Let's compromise it, Sam."

" How ?"

" Let the Frenchman go, and keep his canoe."

" I'm goin' t' keep the Frenchman an' his canoe, too. We may need somebody t' talk the lingo for us."

" But we haven't much left t' eat," protested Jeremiah, earnestly. " We won't be likely t' find anything on our way down the lake, or up either, for the matter o' that."

" There may be worse things 'n that, Jeremiah."

" What ye goin' t' do? Why don't ye say it right out ? "

" I think we'd better take the man with us. We'll make him paddle."

" That's not so bad. We'd better start, then."

" I'll call Tehasen and Totaro, an' we'll do it."

The two Indians came in response to the summons, and reported that no signs of danger had been seen on either side of the place where the landing had been made. The few necessities were carefully apportioned and reloaded, and it was decided that the prisoner should, in his own canoe, follow that of the hunter, while all four were to keep more nearly together than they had done in the earlier part of the night. Additional paddles had been brought, and these were given the two young soldiers and the prisoner; then at the word of Sam, the entire party embarked and glided out upon the lake.

CHAPTER XXIV

WITHIN SIGHT OF THE FLEETS

UNTIL nearly daybreak the canoes continued on their way, passing farther out into the lake when they approached some projecting point, and moving steadily and almost as silently as the silent waters. They kept within hailing distance of one another, however, and as they were now drawing near the place of greatest danger, the precautions of all the men increased.

At last, at a signal from the hunter, the four canoes were drawn to a common point and the leader offered his suggestions as to what should next be done.

"I'm thinkin'," said Sam, "that 'twould be wise for us t' land on the other shore an' carry our canoes overland there. We won't be so likely t' be seen."

"My 'pinion is," spoke up Jeremiah, promptly, "that we'd better land where we be. We can cut in behind th' fort an' it 'll be easier an' nearer t' Champlain."

"Ye're mistaken, Jeremiah."

"No, I'm not mistaken!"

"Well, Jeremiah, 'f you want t' do that, ye can. Seein' as how ye're so set in yer ways there isn't any reasonin' with ye, ye'd better follow yer own ideas an' we'll try the other side. We can meet again anywhere ye say."

"We mustn't separate," protested Jeremiah, sharply.

"No more we should. But that's for you t' say, an' not for me."

"Are you going to the other shore, Sam?"

"Certain, sure."

"Then we'll go with ye, though th' signs all p'int in my way."

"What signs?"

"Jest common sense. Anybody c'n see it's jest as I'm tellin' ye."

The hunter laughed lightly, but said no more, and at once began to send the canoe swiftly toward the shore he had himself preferred. The others followed him and, just as the sun rose, they landed in a little cove. The frail craft were drawn up from the water and concealed within the forest, one of the Indians was stationed as a guard, and the rest of the party assembled about the hunter to eat their breakfast and decide upon the plans for the day. Though the French fort was on the opposite shore some distance below them, their present position was much more perilous than any in which they had yet found themselves.

"Our victuals won't last much longer," suggested Jeremiah.

"They'll have t' last," responded Sam, glibly.

"That's all very well t' say, but what's t' be done if we keep on?"

"Want t' go back to William Henry, Jeremiah?"

"Who said anything about goin' back?" demanded the ranger. "I didn't."

"No more ye did," said Sam, soothingly. "I didn't know but ye'd like to."

"I don't go back after I've once started. You know that as well as I do."

"So I do, Jeremiah, so I do. You're true grit when ye take hold. An' I don't mind ownin' up that we shan't get very fat. We didn't start out for that, though."

"What ye goin' t' do next, Sam? Ye seem t' be havin' everything yer own way."

"I don't stick up for my own notions except when I know they're best," replied Sam, in all soberness. "Now, Jeremiah, don't ye think 'twould be a good scheme for us t' carry our canoes 'round t' Champlain this mornin'?"

"An' not wait till night?" demanded the ranger, in surprise.

"That's it, exactly. Yer mind works like lightnin', doesn't it, Jeremiah, when ye once get it t' workin'?"

"I don't know 'bout that. What's yer plan,

Sam? Ye know, o' course, there'll be a good deal more likelihood o' our bein' seen if we try it in broad daylight."

"That's so. But then, ye mustn't forget that there's a good deal more likelihood o' our bein' seen here where we are, too. I'm thinkin' 'twould be a ticklish job to carry our canoes in the night, an' if we go careful-like in the daytime, we can make up, in part, anyway, for the danger by bein' able t' see what we're up to. That's my idea, but I don't care t' push it if you don't agree with it. Ye know more'n I do 'bout such things 's we're tryin' for to do. What d'ye say?"

"I'm agreed."

"Then we'd better be up an' doin', as the psalm tune says."

The hunter's suggestion was promptly adopted, and as soon as the Indian guard had been recalled, the light canoes were placed upon the shoulders of the men, and the advance through the forest was begun. The prisoner was compelled to carry his own canoe alone, which he did by overturning it and permitting the weight to rest largely upon his head and shoulders, while he held it in place by his hands. The other canoes were held in an upright position, so that the various articles which they contained might be the more easily carried. The two Indians still moved in advance of the little party, because of their greater skill in detecting the pres-

"THE ADVANCE THROUGH THE FOREST WAS BEGUN."

ence of danger, and in silence all seven proceeded on their way.

Frequent halts were made to rest and reconnoitre, but in the course of a few hours the difficult journey had been accomplished, and they were on the borders of the larger lake.

"There," said Sam, with a sigh of relief, when at last the party halted, and the canoes were concealed, "so far, so good. Now if we can do as well for the rest of the way, I guess the colonel won't have much t' complain of. I want ye all now t' take yer blankets an' go in behind those cedars there on our right an' go t' sleep. I'll call ye in time for our start."

"Are we going to eat anything, Sam?" inquired Peter, who was exceedingly hungry.

"Not jest yet. We'll have a bite afore we go, but ye'll save yerselves from bad dreams 'f ye don't try it now. Better keep — "

The hunter ceased abruptly, as one of the Indians, who had withdrawn a few minutes before, returned hastily, and in a manner that showed he had seen something unusual.

"What is it, Totaro?" demanded Sam, in a low voice.

The Indian, instead of speaking, beckoned to him to follow, and the two disappeared. Not one of the remaining party spoke or changed his position, except as he brought his gun into a place where it could be used promptly if occasion demanded it. In breathless silence they waited for the return of their leader, aware that something startling had been seen by the red man. The slow moments passed, but not a sound could be heard save the hoarse cries of crows, passing directly overhead. Through the open spaces among the tall, dark trees occasional glimpses of the waters of the lake were visible, but what had aroused the excitement of Totaro did not appear.

At last, after what seemed hours to Peter, the hunter reappeared, and, motioning for silence and the utmost caution, beckoned them all to follow him. Stepping slowly and carefully, the men advanced until they had gained the shelter of the huge trees near the shore, and then in a whisper Sam said : —

"See that, will ye!" pointing, as he spoke, at the lake before them.

No further explanation was required. Near the opposite shore was seen a fleet of canoes and bateaux filled with white men and Indians. There could be no question as to what their destination was to be, and the little group on the shore watched their movements with an interest too intense for words.

Slowly, steadily, silently, the fleet moved forward, the men wielding their paddles deftly, and apparently eager for the task before them. There must be two hundred at least in the force, Peter thought, as he excitedly watched their movements. He could hear the heavy breathing of John, who was standing by his side, and required no further information as to the excitement under which he was laboring. The prisoner, who had been compelled to come with them, was gazing at the fleet, and the light in his eyes, as well as the smile on his face, betrayed the exultation which was his as he watched the movements of his friends. Peter trembled as he thought what the result of a shout from the prisoner would be, but the man plainly was not minded to incur the danger which would accompany such an act on his part, though doubtless his hope of rescue was now strong.

At last the long procession passed from sight, and after he had left one of the Indians to watch and report instantly anything that he might see, Sam bade his companions return with him to the place where they had previously been.

"Looks pretty bad, doesn't it, Jeremiah?" he inquired, as he seated himself on the end of a log.

"Does that!" assented the ranger. "Th' signs are all agin us."

"How's that? Isn't this jest what we came for to see?"

"I s'pose 'tis. Goin' back t' th' fort, now?"

"Goin' back? Why, Jeremiah, we aren't half-way, yet."

"What are we goin' t' do for something t' eat?"

"Have t' get it, I s'pose," replied the hunter, glibly. "We've got some here, an' we'll have t' be on the lookout for more."

"Have t' do more'n 'look out,' I'm thinkin', afore we're back at Fort William Henry," muttered the ranger.

"Jeremiah, what ye need now is a nap. Ye'll feel better when ye wake up. Now you and all th' rest o' ye take yer blankets an' go t' sleep. I'll keep watch, an' if yer presence is required, I'll let ye know."

"You lie down an' I'll do th' watchin'," suggested Jeremiah.

But the hunter was not to be turned from his purpose, so the other members of the party withdrew into the thicket, and wrapping themselves in their blankets were soon asleep.

At sunset the hunter roused them, and a scanty supper was served. It was not deemed wise or safe to have a fire, and when at last the canoes were again placed in the water Peter was rejoiced, for the night air was cold, and he was eager for the task which would serve to keep him warm. The hunter had reported that there had been no sign of the presence of danger, nor had Tehasen discovered any more passing canoes. But the sight which all had seen had served to inform them of the necessity for watchfulness, and no warning words from Sam were required to make every man cautious and careful.

Keeping well together, the four canoes were soon passing down the lake, but when, near morning, a landing was made, no further evidences of the presence of Frenchmen or Indians had been discovered.

On the following day, another fleet of at least a hundred canoes and bateaux was seen passing up the lake in the direction of Ticonderoga. It needed no special powers of prophecy to predict that the enemy was assembling, and that within the near future something would be done that would call forth all the energies of the colonials stationed at Fort Edward or Fort William Henry.

On the fourth morning, when the party landed, Jeremiah was vigorous in his protests against a further advance. " We've seen enough, an' we haven't got victuals enough to last us more'n two days more. I say," he firmly declared, " that th' thing for us t' do is t' go back an' report. We've seen what ought t' make both General Webb at Fort Edward an' Colonel Monro at Fort William Henry know that they've got t' get ready, an' right away, too."

" We'll go on one more night, an' then we'll turn back if ye say so," said Sam, quietly.

" Turnin' back an' gettin' back is two different things," muttered Jeremiah.

" One night more, an' that 'll be th' last.'

"May be th' last in more ways 'n one, unless all the signs are wrong."

The hunter laughed good-naturedly, but was not to be turned from his purpose. He had received a special word from Colonel Monro which he had not explained to any of his companions, and this it was which was leading him on. He had been hoping that the project for which he had come might be accomplished in the preceding night, but although he had maintained an unusually careful watch, he had not discovered signs that he was near the place he was seeking.

On this night, however, the canoes had not been on the lake more than an hour before his keen eyes perceived a light near the shore in advance of them, a fact that instantly caused him to give the signal for all to approach the place where his own canoe was resting on the water.

"We must land right here," he said, "or rather I want you all to go close inshore an' wait for me. I may be gone a long time, an' then again I mayn't. That's as may be."

"Where are you going, Sam?" inquired Peter, in a low voice.

"Up ahead."

"Going alone?"

"No, Tehasen is goin' with me. We'll go ashore an' change canoes, for I don't want t' take any chances out here."

"Don't you want me to go with you, Sam?" said Peter, eagerly.

"No, I don't; leastwise, I don't as yet. Maybe I shall a bit later; we'll see. Now then, go straight for shore, an' don't make any noise, either."

Silently the canoes were paddled to the shore, and there the transfer was made, Tehasen taking his place with the hunter, and Jeremiah joining the other Indian. Still Sam was silent concerning his project, but Peter understood from his actions that something of supreme interest was about to occur.

"Better wait right here in yer canoes," suggested the hunter as he prepared to depart. "It may be ye'll want 'em in a hurry. Now then, Tehasen," he added, turning, as he spoke, to his Indian companion.

Silently and together the two paddles were dipped into the water, and the canoe glided out into the darkness, leaving the men near the shore completely mystified and fearful alike for their departing companions and for themselves.

CHAPTER XXV

The Fire on the Shore

MORE than two hours had passed before the hunter and Tehasen rejoined their waiting companions. Thin clouds had been gathering across the sky and had shut out the light of the stars, and when the little craft silently glided amongst the three waiting canoes, its sudden appearance was so startling that, in spite of the fact that all four of the men had been anxiously awaiting its coming, Peter could scarcely repress the exclamation that rose to his lips.

"We've found it," said the hunter, in a low voice, as the speed of his canoe was checked.

"Found what, Sam?" inquired Peter, eagerly.

"What we came fer t' see."

"What's that?"

"Th' place where a lot o' their stuff is stored. The colonel had word o' it, but he wanted t' make sure, so he sent us."

"And that was what we came for? Why didn't you tell us, Sam?"

"No use in tellin'. Time enough to tell after we'd found it. If we hadn't found it, no one would have been disappointed, an' no harm done."

" You're sure about it now, are you, Sam ? "

" Never more certain o' anything in my life. Tehasen an' I landed an' crep' up near the spot, an' there isn't a handful o' men there, either. I'm thinkin' this Frenchman — what d'ye say his name is ? "

" Montcalm ? "

" Yes, that's it, Montcalm. I'm certain sure he must 'a' been mighty careless t' leave things that way. He prob'ly never thought as how the colonel would ever send any o' his men clear up here t' take a look at it."

" What's t' be done now, Sam ? Do we turn back now ? "

" I wish we could get some o' th' victuals there," suggested Jeremiah.

" We c'n do the next best thing t' gettin' 'em fer ourselves," said the hunter.

" What's that ? " inquired the ranger.

" Burn 'em up so that the Frenchmen can't have 'em fer themselves."

" How's that t' be done ? " said Jeremiah, quickly.

" There's only one way t' do it."

" And that's ? "

" Jest t' do it."

" How ? " persisted the ranger.

" I'll tell ye. My plan is fer us all to go down the shore an' land right near the spot. Then Tehasen an' Peter an' I'll creep up an' set fire t' the place from two or three sides at the same time. As I told ye, there aren't very many men there t' guard th' stuff, an' jest 's soon 's it's on fire they'll have enough t' do t' fight th' blaze, to say nothin' o' us."

"Ye're sure it can be done, Sam?" inquired Jeremiah, thoughtfully.

"I'm sure it c'n be tried. That's what we come fer, to try, wasn't it?"

"I s'pose so. Don't ye think I'd better go 'long with ye 'stead o' Peter?"

"No, I don't. We may fail, an' then again we mayn't. That's as may be. But we've got t' have somebody on th' shore waitin' fer us that knows jest what t' do, an' you're the man for that, Jeremiah. There's no mistake 'bout that. See what I mean?"

"Yes, I s'pose so," admitted Jeremiah, reluctantly. "I'll do 's ye say, Sam."

"But don't ye think that's th' best thing t' do?" persisted the hunter.

"It 'll do as a starter. If ye get in any trouble, though, ye must call out for us. We shan't leave ye, though I'm thinkin' ye know that already."

"That's th' very thing I mean, Jeremiah," said Sam, earnestly. "I'll take charge o' th' settin' fire t' th' stuff an' you'll be the one t' be on th' lookout for us, an' ye'll know jest what t' do. There

isn't 'nother man in the whole o' Webb's army c'n
do that 's you can, Jeremiah.''

"Why don't ye set about it, then? It 'll be
mornin' 'fore ye get started."

"I'm ready. I jest wanted t' have everything
understood. Peter an' John would do better t'
keep close t' our canoe."

"We all must keep together, Sam," said Jeremiah,
seriously.

"So we must, so we must; but I was only
'rangin' some o' th' details, ye see. I was jest
suggestin' th' order o' the thing."

"What 'll we do with this prisoner?" inquired
Jeremiah, glancing at the man they had taken, who
was sitting silent in his canoe near them.

"You'll have t' keep an eye on him yerself,
Jeremiah," responded Sam. "I'm not afraid o' anything happenin' t' him 's long 's he's in your charge."

There was a further low conversation, in which
some of the minor details were provided for, and
then the four canoes glided out from their place of
shelter, Sam and Tehasen leading the way, and
Peter and John following closely, while the other
two were but a short distance behind.

Peter was intensely excited, but his movements
were cautious and he did not speak to his companion as they deftly wielded their paddles and
kept close to the canoe in advance of them. In
spite of the hunter's calm manner he was convinced

that Sam looked upon the venture as one of great
peril, and that he was not at all confident as to
its outcome.   The importance of the attempt they
were about to make, however, was so apparent,
and the eagerness of the hunter was so keen, that
Peter found himself sharing in the excitement; and
when at last, at a signal from Sam, the canoes were
turned sharply to the shore and drawn up on the
wooded bank, he was almost as impatient as his
friend for the advance to be made.

There was another brief, whispered consultation,
then the hunter and Tehasen and Peter left their
companions on the shore and entered the sombre
forest.   The three men kept close together until
they arrived at a spot from which they could get
a view of the place where the stores were kept.
There was a rough building of logs and bark, and
outside and near it were many piles of various
articles of food or supplies which the oncoming
army would require.   Not a man was in sight, but
Peter knew as well as his comrades that that fact
did not imply that no guards were near.

At a word from Sam, Tehasen darted silently
into the forest, leaving the hunter and Peter to
await his return.   "He'll find out th' lay o' th'
land," whispered Sam, and Peter nodded his head
as a token that he understood.   The darkness of
the night had deepened, and all things seemed to
favor the undertaking.

The Indian returned with the report that no sentries had been stationed, or that he had not been able to perceive any, and if guards were there, they either were within the rough building or were sleeping, in their confidence that no danger was to be feared from prowling provincial soldiers.

"So far so good," whispered Sam. "Now, then, we'll creep up an' set fire t' th' stuff on three sides at the same time. Peter, you take th' side nearest us, the one we c'n see right before us, an' Tehasen an' I'll creep round t' th' other side. Wait here a bit before ye start so 's t' give us a chance t' find our places before ye begin, but don't wait too long; an' jest 's soon 's ye strike a spark an' think th' fire is goin' t' go, then light out for the shore an' don't wait for us. Tehasen an' I'll have to take care o' ourselves."

"Shall I wait for you where the canoes are?" whispered Peter.

"That's as may be. Ye know we 'greed to make for that place where ye waited for us before, an' then t' wait a spell till all o' us could get together. But ye may not be able t' do that. Everybody 'll have t' look out for himself, an' it may turn out that we shan't see one another till we're all back in Fort William Henry. But I'll trust ye, Peter. Do yer best, an' I'm not afraid fer ye."

Without another word the hunter and his Indian

companion turned into the forest and Peter was left to himself. He gazed earnestly at the place he was to seek, trying to select the best course to pursue. The stores were on a high bank or bluff, and could be seen from the open lake, for they were only in part within the shelter of the forest. It certainly was plain that the Frenchmen had neglected even the ordinary precautions, in their confidence that the spot was so far from their enemies that they had nothing to fear for its safety. And yet, to the excited young soldier the suggestion occurred that the stores were better protected than appeared to a casual observer. Men might be in waiting within the low building, and even now keen eyes might be watching the movements of his own companions. The thought was not reassuring, and in the increasing excitement which it produced, Peter decided to begin his own attempt. He felt again in his pocket to make certain that his flint and tinder were secure, then, drawing a long breath, moved swiftly toward the border of the forest. The great trees were growing within a few yards of the place he was seeking, and he swiftly darted from one to another until he had gained those nearest the building. He was now in a fever of excitement, though his actions were still carefully made. His breathing was short and rapid, and he was looking before him with an intensity that was almost painful.

At last, assured that no enemy was in sight, he quickly crouched low and ran across the few yards of clearing that lay between him and the side of the building. In a moment he had gained shelter, and flinging himself upon the ground, he listened in an agony of suspense, to discover if his presence was known.

But not a sound broke the tense silence of the night. The tall, spectral trees were motionless, and the clouds in the sky almost appeared to stop in their course as if they, too, were interested spectators of the scene below them. After a little Peter recovered a measure of self-possession and drew forth his flint and tinder. He wondered if his two comrades had been as successful as he, and were now ready for the final attempt to be made. There was no way in which to ascertain that, and no time to lose. He felt of the side of the building before him. It was rough, and the portion near him, anyway, was covered with bark, the very thing he most desired.

With trembling hands he struck a spark, and to his delight the very first one caught in the tinder and a tiny flame began to creep up the bark. In a moment the flame increased, and, convinced that his attempt had been successful, he rose to his feet, unmindful of everything except the necessity of fleeing from the spot, ran swiftly across the little clearing, and gained the protection of the forest.

There he paused for a moment and waited to learn how it had fared with his friends.

The smoke and fire from the blaze he had kindled were increasing, and suddenly his heart leaped within him as, from another side of the building, he saw a curl of smoke rising in the dim light. The hunter, too, had been successful, and at the sight Peter felt a wild exultation sweeping over him. But why did not Sam or Tehasen appear? In his eagerness he had forgotten the careful instructions which the hunter had given him, and still stood intently watching for the coming of his friends. The smoke now was pouring in streams from the building, and there was a great shouting within. Men rushed from the structure, and then, after a brief moment of confusion, a line was formed and buckets of water were passed from hand to hand, until their contents could be dashed upon the blazing bark and logs. There were sharp calls and quick commands, and Peter was amazed to perceive how many men were answering to the summons. The hunter must have been mistaken, he thought, in his opinion that only a few were in the place, for now it seemed as if scores were to be seen running about in all directions and fighting the flames.

He was recalled to the necessity for action on his own part, when he saw that out of the apparent disorder a systematic movement was coming.

He became aware not only that some of the men were striving to put out the fires, but that others were intent on other purposes. At the call of the officers, certain ones darted toward the shore, others formed in bands and advanced toward the forest, and some were running toward the place where he himself was standing. He turned and began to run for the spot where the canoes were in waiting.

Unmindful now of caution, disregarding the sound of the fallen branches that snapped under his feet, he was filled with the one thought of gaining the safety which the canoes could give. Behind him he could hear the sounds of the advancing men, and the fact that in the number he had seen some of the hideously painted savages whose faces, daubed with yellow and vermilion, shone fiendishly in the light of the fire, increased his wild desire to escape. Leaping from mound to mound, turning neither to the right nor left, he gained the shore and looked before him for the friendly canoes. Not one could he see.

Amazed and terrified now, he halted for an instant, then quickly deciding that he had made a mistake and had not returned to the place where he had left his friends, he began to run still more swiftly up the shore. Even while he was increasing his speed he could see that the trees and bushes appeared familiar; but the canoes certainly were not there, and in his wild desire to escape he still

sped aimlessly on and on.   The knowledge that
white men and red were behind him, and not far
away at that, made him redouble his efforts, though
he had no plan in his mind, and no thought of anything now save escaping in some way.

CHAPTER XXVI

The Flight in the Darkness

PETER was again close to the border of the lake, and as he neared the bank he perceived a canoe lying close inshore. It held a man whom, in the dim light, he thought he recognized as John. Thinking only of the peril from those behind, he darted into the open space, and in a low voice called, "John! John!"

"Is that you, Peter?" was the response, given in a guarded whisper.

"Yes, yes! Take me on board! Be quick!"

The canoe was instantly driven to the shore, and in a moment Peter was on board and had seized a paddle.

"Now, John, with all your might! Paddle as you never did before!"

"Are they after us?" inquired John, as he obeyed.

"Yes, yes! And close behind, too!"

Not another word was spoken as the two young soldiers, with their long and powerful strokes, sent the light little craft away from the shore and headed directly toward the waters of the open lake. Peter was watching the place where he had embarked, momentarily expecting to perceive the forms of his pursuers or to hear the report of their guns. Every second was precious, and if only they could gain a good distance from the shore, they might escape. His companion required no urging, for he fully comprehended the peril, although there had been no opportunity for explanations. Together their paddles struck the water, and the canoe darted forward under the urging of their united powers. The increasing darkness would favor them, they both thought, and their one purpose was to place the greatest possible distance between themselves and the shore.

Suddenly from the lake below them came the loud report of a rifle, and for an instant the flash of the gun could be seen in the darkness. Moved as by a common instinct, both young men lifted their paddles from the water, and Peter whispered, "It's a canoe, John, and straight below us."

"What 'll we do?"

"Put straight up the shore. It's our only hope."

The canoe was headed up the lake, and its speed was redoubled, as the young men realized that their greatest danger was from pursuers who were embarked as they themselves were. The darkness was so dense that they could not obtain even a glimpse of their enemies, but the shot, though it had gone wide of its mark, had indicated only too plainly what and where the danger was.

Speed was the one requisite, and all precautions
for silence were abandoned, although both Peter
and John were skilled in the use of the paddle, and
an ordinary observer would scarcely have detected
any sound as their blades struck the water together.
But both young men knew that they were not pursued by amateurs, and the very fact that they had
been fired upon, from a direction where neither
had suspected danger, had its unmistakable lesson.

The fear of a repetition of the shot was strong,
and for a time served to inspire them to do their
utmost. The dim outlines of the wooded shores
flew past them; the bow of their little canoe was
lifted with every stroke of the paddles. Perspiration was soon running in streams down their faces,
though neither was aware of the other's condition.
It was a race for life and liberty, and the prize was
worthy of their greatest endeavors. The fact that
the shot had not been repeated afforded a measure
of comfort, but would not permit either to relax
his efforts.

On and still on they swept, until at last what
Peter considered a full hour must have passed.
The strain of the continued paddling was beginning to tell, and arms and back and shoulders were
numb and aching. Nothing more had been seen or
heard of their enemies on the water, and Peter was
convinced that they must long since have distanced
any who might have been running along the shore.

Flesh and blood could no longer endure the terrible strain, and, lifting his paddle from the water, Peter said in a hoarse whisper to his companion : —

"Hold up a bit, John.   Let's get our breath and see what's going on."

Without replying, John did as he was told, and both boys listened intently as they tried to pierce the darkness behind them.   The canoe was still gliding forward, but was as silent as the dark waters.   The clouds overhead were darker still, and a raindrop fell upon Peter's upturned face.

"Good!" he muttered.   "We're out of it all right now."

"S-s-sh!" whispered John, warningly.

A faint sound could be heard for a moment in the direction from which they had come, but it was only for a moment, and then the deep silence again rested over all.   It was, however, suggestive, and turning the canoe partly about, both boys listened even more intently than before.

The rain had begun to fall, a spring shower, gentle and quiet, and as the drops touched Peter's hot, wet face they were marvellously cooling and refreshing, but only in a general way was he mindful of it.   Eagerly listening for the repetition of the sound that he had heard, he was almost unaware of his immediate surroundings.

"Nothing!" whispered John.   "Shall we go ashore or go on?"

"What time do you think it is, John?"

"After midnight."

"No doubt about that.   I'm dead tired."

"So am I.   We don't know where to land if we try to go ashore."

"We can find a place all right."

"Why don't we stay right where we are for a while?   This rain won't spoil our clothes, and we can rest as well here as anywhere, can't we?"

"Yes, if —"

Peter stopped abruptly as again the report of a rifle was heard, and this time a wild shout accompanied it, which to the startled young men seemed to come from only a short distance behind them. The aim of the marksman had been better, too, or chance had favored him, for there was a sudden movement of the canoe that indicated it had been struck.   It was so dark that neither of the boys saw that the bullet had passed directly through the sides of the canoe, leaving a hole on either side just above the water.

"Hit, Peter?" whispered John, as he grasped his paddle again.

"No.   You?" replied Peter.

"Sound 's a whistle," responded John; and silence followed as the canoe was driven swiftly forward, heading again for the waters of the open lake.

The desperate attempt to shake off their pursuers was renewed, and without thought of the direction in which they were moving, the two fugitives endeavored to gain some place where they might escape from those who were so persistently following them.

For a time their efforts seemed to be successful. They did not hear again the report of the gun, nor was the one loud shout repeated. The falling rain and the darkness provided their strongest hope of defence, and after they had been sweeping over the water for what seemed to Peter to be sufficient time to distance the canoes that were following them, he said : —

"John, let us double on our tracks."

"Go ahead, I don't care; I'm almost used up, anyway."

"What's this water in the canoe, John?"

"I don't know.   Rain, I s'pose."

"It's more than rain.   I'm afraid the canoe is leaking."   As Peter spoke he changed his position slightly, and the increased weight instantly brought the holes which had been made by the bullet below the level of the water.   The quick ears of the young provincial detected the sound of incoming water, and he speedily discovered the source of danger.

"It is leaking, John," he whispered.   "There's a hole on either side.   Wait till I try to plug them up."

Drawing from his pocket a small strip of linen cloth, which the hunter had insisted should be brought when they had set forth from the fort, — for no one knew better than he the need of such an article if one of the party should chance to be injured, — Peter tore away a portion and rolling it up hastily pushed it into the small rent that had been made by the bullet. The hole on the opposite side of the canoe was treated in a similar manner, but by this time the water in the little craft was at least two inches deep.

"Can we bail her out?" whispered John, anxiously.

"I'm afraid we'd be heard. We'd have to use our hands, anyway. I think we'd better put straight for the shore," replied Peter, in a whisper.

"Think it's safe?"

"Safer than to be out here in a leaking canoe. I think there's a good chance that the Frenchmen have gone on, and we can just slip ashore behind them. We'd better start, or it 'll be morning and we'll be caught then, anyway. Give way, John."

The whispered conversation ceased, and again the young provincials began to send the canoe in toward the shore. Their progress now, however, was much slower, and they were fearful for the safety of their little craft as well as of their own peril from their enemies. Frequent stops to listen were made, but not an alarming sound could be heard.

The rain had ceased, and the low clouds were break-ing. From the shore came the odor of trees upon which the rain has been falling in the warm days of late spring. There was a warmth in the air indicative of still warmer days to come. Under other circumstances Peter and John would have been among the first to respond to these quiet influences, but at the present time the supreme hope in their minds was to elude their persistent pursuers, and gain the shore before the canoe should fill and sink. The water in it was steadily growing deeper, and the little craft was every moment becoming more unwieldy, and rolled with the slightest movement of the men on board.

"A couple of hundred feet more and we're done," whispered Peter, encouragingly.

"If we can ever make it," replied John, in a hoarse whisper.

"We've got to make it! It's our only chance."

John did not reply, but as he looked for a moment up the lake, in the dim light he perceived what he knew must be an approaching canoe. The sight was appalling, for if he could see the other canoe, it was but fair to conclude that their own could also be seen.

Turning sharply about he began in a whisper, "Peter, there's a — "

The sentence was not completed, for with a sudden and unexpected lurch the canoe overturned, and both occupants were thrown into the water. In a moment Peter rose to the surface and looked about for his friend. He perceived John's head above the water, and hastily swimming toward him whispered, "Can you make the shore, John?"

"Yes, I think 1 can."

"Come on, then. We'll let the canoe go. If they find it, they'll think we've been capsized. Keep close to me, and put your right hand on my shoulder when you are tired. It's not very far, anyway."

Peter was an expert swimmer, and John was an unusually good one, also, but the exertions of the night and the long abstinence from food had combined to make him well-nigh helpless. Peter, swimming beside his friend, watched him anxiously. They had covered successfully half the distance that lay between the place of their accident and the shore, when suddenly a fierce shout came from the lake behind them. Peter knew the cause of it and who his pursuers had been, for doubtless the discovery of the overturned canoe had given rise to the exultant yell, and no one but the red men could ever voice such a wild and terrifying cry.

Instinctively he began to increase the speed at which he was moving, but quickly realizing the weakness of his companion, repressed himself and came close to John's side. He saw that his friend was beginning to flounder and that his efforts to keep himself afloat must certainly draw the attention of their keen, observant foes.

"Put your hand on my shoulder, John," he whispered; but he was compelled to assist before the simple act could be done.

Then slowly and painfully Peter continued his way to the shore. There were moments when it seemed to him that he must abandon the attempt, when the surrounding darkness seemed to give way to a darkness denser still, when his breath came in gasps, and there was a pain in his side like the sharp cutting of a knife. But still he struggled on, slowly nearing the goal; and at last, stumbling and falling, the two young provincials gained the longed-for refuge. After casting one keen glance behind them to see if they were followed, they crept within the shelter of the forest, where, despite their wet condition, they cast themselves in utter weariness upon the ground and were speedily asleep.

The sun was shining brightly when they awoke; but it was not the glory of the morning that drew their attention, for directly before them was a sight which banished all thoughts of other things from their minds.

CHAPTER XXVII

FATHER ROUBAUD

STANDING directly in front of them and looking down with an expression by no means
hostile, was a large man, clad in priestly garments,
and evidently as greatly surprised as were the
youths themselves. A multitude of thoughts and
fears flashed into Peter's mind as he gazed in
silence at the man before them. Who was he, and
where had he come from? Where one white man
was, it was probable that others also were to be
found. Was there a camp near by? Did this
unexpected presence imply that at the very time
other men were also watching them? The thought
was not comforting, and he glanced anxiously about
him, an action that brought a reassuring smile to
the face of the man.

Peter and John rose to their feet and approached
the stranger, who still looked kindly at them and
did not move as they came near.

" Where are we ? " demanded Peter, anxiously.

The priest smiled in reply, but when he spoke,
his words could not be understood.

"He's a Frenchman," said Peter, in a low voice, to his comrade.

"So much the worse for us, then," replied John, soberly.

"We'll have to wait and see. He seems to be friendly."

"There's only one kind of a Frenchman in all the world."

The man was speaking again, but his words were still unintelligible. Only one word was repeated several times, the word "Roubaud," but it conveyed no meaning to either of the young soldiers, both of whom were ignorant of the French language and could not know that this was the name of one of the best and most justly celebrated of all the priests who were accompanying the expedition of Montcalm. He was a man of large sympathies and of a most tender heart, and his good works were well known throughout the advancing army. Much of his time had been given to labor among the red men, particularly among the so-called "Mission Indians"; but even if Peter and John had known all this, it is to be questioned if their hearts would have been any more at rest. As it was, it was simply evident that he did not understand them when they spoke, nor they him; but his presence was certainly startling, and to the boys seemed to promise little good.

Further attempts at conversation were abandoned

when it was evidently impossible for either to make the other comprehend his words, and then the priest with a kindly gesture beckoned for the young soldiers to follow him.

" I don't like it," said John, in a low voice.

" Neither do I, but I think we'd better go with him," replied Peter.

" We don't know where he's taking us."

" We'll have to take our chances. If he has men near by, there's no use in trying to get away. He seems to be friendly."

" But he's a Frenchman," protested John, as if the declaration was sufficient.

" We'd better go, John."

The priest, who had been listening to the low conversation, again began to speak rapidly, and to try to show that his purpose was not hostile. Several times he pointed toward some place down the shore which the boys could not see, and at last, though with evident reluctance, both nodded their heads to indicate that they were disposed to do as he had suggested.

The face of the priest beamed with satisfaction, and he began to walk in the direction he had indicated, the two young men following slowly and glancing keenly about them as they proceeded. Their suspicions moved the man to continue his friendly gestures, but the fears of his followers were by no means relieved when, in a brief time, they arrived

at a place from which they could see an Indian
camp or village extended before them.   It was too
late to turn back now, for their presence had been
discovered.   Women and children noisily advanced
to meet them, plainly with no friendly purpose;
but a wave of the priest's hand caused them all to
fall back, though they stood and watched the little
party, their dark eyes glowing with hatred, and the
expressions upon their faces boding little good for
the young prisoners, for such both Peter and John
now thought themselves to be.

Peter glanced for a moment at John and saw
that his face was deadly pale.   He wondered if his
own face was of the same hue, for certainly the
fear in his heart was as great as that in his companion's.   Not a word was spoken as the priest led
them to an unoccupied hut or wigwam, and holding
aside the blanket which served as a door, indicated
that they were to enter.

There was nothing to be done but to obey, so
both boys stepped inside and at a gesture from the
priest seated themselves upon the ground.   Their
conductor now spoke earnestly to them, but his
words conveyed no meaning, though the tone was
still friendly; and after a few moments he departed,
dropping the blanket into its place as he left the
hut.

Once more by themselves, John was the first to
speak.   "We're in for it now, Peter," he said dis-

consolately, "and we'll never see Fort William Henry again."

"Can't tell about that. As Sam says, 'That's as may be.' But we're all right so far, and we'll have to take things as they come. It 'll be time enough to give up when we have to."

"Have to now, I'm thinking," replied John, greatly depressed.

"The man didn't seem to be so very bad," suggested Peter.

"The man? I'm not thinking of him. There are hundreds of redskins here, and what they've come for you know 's well 's I do. What fools we were ever to follow him! We've walked straight into the trap."

"I'm not so sure o' that. It may come out all right, yet."

John shook his head and became silent. The noise of the people in the village as it came to their ears was by no means reassuring. Wearied, hungry, wretched, the young prisoners certainly did not have a pleasing prospect; but Peter, who was the more resolute of the two, was striving to think out what it all meant. The apparent friendliness of the priest, the fact that his influence among the savages had been sufficient to save them from insult or injury thus far, and that he had led them to a place where they at least were not molested, were all in their favor, though no one could deter-

mine what still remained to be faced or endured.
At that very moment the blanket was partly drawn
aside and the faces of several Indian boys could be
seen peering in at them.

"Get away from there!" shouted John, savagely,
springing up and looking about him for some
missile to throw at the visitors.

The blanket was instantly dropped, and the boys
fled, though, unknown to John, the cause was to
be found in the approach of the priest rather than
in his own threatening words or actions.

Father Roubaud now entered, bearing in his
hands some food which he placed before Peter and
John, motioning for them to eat it.   With a glance
at his companion Peter instantly responded to the
invitation, and John soon followed his example,
the priest, meanwhile, standing near, his genial face
indicating that the pleasure of the repast was not
all theirs.

When they had finished and the priest prepared
to withdraw, he held back the blanket and indicated that the prisoners might accompany him;
but both shook their heads and declined.   It was
better to endure the silence of the hut than the
noisy demonstrations which their appearance would
at once arouse in the village; but when their visitor
had withdrawn, and the young men had resumed
their seats on the ground, the situation did not look
quite so gloomy to either.   For some time they

talked over their predicament, but at last concluded that it was both wiser and safer for them to remain quietly where they were than to venture forth among the savages.

"And yet he seemed to mean that we could go out if we wanted to," said John.

"I think so. Perhaps we'll try it by and by. Hark! What's that?" he said abruptly, rising as he spoke.

There was slight need of bidding his companion to listen. A wild, prolonged shout rose from the midst of the village, and the shrill cries of boys and women were mingled with the deeper tones of the men. People were running about, and as the moments passed, the excitement, instead of subsiding, rapidly increased.

"What's that, Peter?" demanded John, sharply.

"You know as much about it as I do," replied Peter.

"They're going past us," said John, after a brief pause.

"Come on, John! I'm going to see what it is, anyway. They can't do any more than turn us back. It may be our chance. Come on! Come on!"

The two young men stepped forth from the hut. All about them the warriors and women and children were running and shouting, and all were moving in one direction, which was toward the near-by shore of the lake.

"Come on, John. We'll go 'long with them," said Peter, in a low voice. "They don't seem to be paying any attention to us."

The two prisoners followed the wild rabble until they reached a spot which revealed the cause of the excitement. Approaching from the lake was a fleet of twenty or more canoes, each occupied by two or more Indians, while in one of the larger canoes in advance were seated three or four white men, evidently prisoners, whose coming had aroused the excitement in the little village.

Before the canoes were beached, the savages on shore darted for the forest, from which their wild yelling could still be heard, and came leaping and running back, each one holding in his hands a stout club.

"I know what that means," said John, in a low voice.

"Yes, the gantlet," replied Peter, his voice trembling as he spoke.

"Horrible! Horrible!" muttered John; but neither turned from the spot, the very horror of the sight they were about to witness holding each in his place.

By the time the howling mob had returned to the shore, the canoes were ready to be drawn up on the land. As the prisoners stepped forth it was seen that about the neck of each was tied the end of a long leathern cord or strap, the other end being securely held by the captors. Peter had been gaz-

ing earnestly at the unfortunate captives, fearful of
discovering the hunter or some of his other friends
among the number. It was a matter of some relief
to find that he did not know any of the wretched
men, but the expression of terror on their faces,
wet with great drops of perspiration, made him
shiver with helpless sympathy.

The greater part of the howling mob were now
drawn up in two lines, facing each other, and, without the formality of a trial or discussion such as
every prisoner was supposed to be entitled to in
times of peace or war, one of the captives was led
to the lower end of the two lines, the strap was
removed from his neck, and he was pushed in between the lines.

Perhaps the wretched man was stiff and sore
from the position he had been compelled to maintain in the canoe, or he may have been wounded
before being made prisoner, for he stumbled and
fell under the blows that were rained upon him
before he had run six yards. With shouts of deri-
sion he was lifted to his feet and again compelled
to run, but after a few steps he again fell, and this
time did not rise.

With louder and wilder shouts the savages turned
to demand another victim, but before the strap had
been taken from the neck of the second unfortunate Father Roubaud stepped forth, and at his
word and uplifted arm a hush fell over all.

Glances of disgust or anger flashed from one dusky face to another, but the hold of the priest upon his audience was strong, and they remained silent, as he earnestly addressed them. Peter did not understand a word that was spoken, but the eloquence and earnestness of the priest were apparent. For several minutes he spoke, and the tones of his voice indicated that he was pleading with them to spare the prisoners. Sullen looks spread over the faces of his hearers, and murmurs of dissent began to be heard when several men, who were plainly chiefs, stepped toward Father Roubaud.

A low and earnest conversation followed, but not even the people in the lines could hear what was said, although their interest in its outcome was as keen as was that of Peter and John, to say nothing of the prisoners. After a momentary hesitation on the part of the chiefs, they turned and spoke a few words to their followers. There were many glowering faces to be seen as the mob slowly dispersed, but the remaining captives were speedily borne away and the excitement was at an end. Peter never knew what it was the priest had said, nor that he had promised if the lives of the prisoners should be spared and the unfortunate men sent to Montreal the reward of their captors should be doubled.

"Looks a bit better for us," suggested John. "The redskins do what that man tells 'em."

"Yes, perhaps so; but I wish we were out of it."

"Why don't we make a break, then? Doesn't anybody 'pear to be paying any attention to us."

But when the two young provincials turned away, they speedily discovered that John's words were not true, and that they were by no means left unguarded by the red men of the village.

CHAPTER XXVIII

In the Indian Village

ALMOST unaware of the direction they were following, Peter and John had approached the border of the village. Before them stretched the great forest, its depths promising shelter and protection from the savages and such terrible scenes as those which they had just witnessed.

"Let's make a break for it!" said John, in a low voice, unconsciously expressing the thought which was also in the mind of his companion. Anything would be better than to remain where they were, and perhaps be the next upon whom the excited red men might turn.

As John spoke, Peter studied the forest before them. Its very silence seemed to summon them, and within its depths safety might be found. Behind them lay the Indian village, where the shouts and cries had not yet been entirely silenced. The temptation to follow his friend's suggestion was strong, and almost as if he were speaking to himself, Peter said : —

"It's the best thing to do. There's no hope here."

"Come on, then! What are we waiting for?" responded John, preparing to run.

"I'm agreed. We'd better not run at first, though. We can keep on slowly, and when we're a little farther from the village we can —"

Peter stopped abruptly, as directly behind him he heard the shouts and calls of a small party of the red men who were running toward them.

To escape now was not to be thought of, and Peter promptly said: "No use to try it, John. They're after us, and we'll have to wait till some other time."

"We'll never make it! We're in for it, Peter," said John, disconsolately.

"Don't give up, John. We'll keep our eyes open and try it again, and perhaps we'll have better luck next time. Don't let them see that we've had any thought of trying to get away."

The half-dozen Indians had placed themselves between the two young men and the forest, and brandishing their clubs they spoke savagely, showing by their gestures that Peter and John were to turn back into the village. The words could not be understood, but the actions were unmistakable, and both retraced their way, striving to appear calm and to conceal the fears that had again seized upon them. Their ready and prompt response appeared to satisfy the savages, whose threatening actions abruptly ceased, and who, to all appearances, now ignored both captives. But the young provincials clearly understood that though the immediate danger was past they still were being closely watched, and that any suspicious act on their part would instantly bring upon them the attentions which they were most desirous of avoiding.

"We seem to be all right as long as we keep in the village," suggested Peter.

"That's where we'll have to stay for a while, then," replied John.

"We can stand that if we have to."

"Yes; but what are they keeping us for? That's what troubles me most of all."

"No one knows. If they'd meant to treat us as they did those prisoners, we'd have known it before this."

"I'm not so sure about that."

"Then, too, you mustn't forget 'twas the priest who brought us into the village. He seems to have a pretty good hold on them."

"What do you suppose he brought us here for, Peter?"

"It may have been to keep the redskins from getting us. Anyway, I'd a good sight rather be his prisoner than a prisoner of these savages."

"So would I. But he hasn't been near us lately."

"That may be all the better for us."

"I don't see how. Peter, don't you think we'd better go back to the wigwam?"

"Yes. We'll attract less attention there, and just at the present time that's what I want more than anything else. We don't want to be too popular."

Accordingly, the two turned toward their wigwam, which lay on the opposite side of the village. Within its shelter they would be free from the attentions of the Indian boys and could talk over their plans and possibly devise some scheme of escape. They were walking near the outskirts of the place, striving to avoid the groups where the boys and women were congregated; but as they drew near their wigwam they stopped abruptly, for directly before them they beheld a dozen or more of the savages assembled about a fire over which hung a great kettle.

"Peter! Peter! Do you see what they're doing?" demanded John, in a voice of horror, and grasping his companion by the arm as he spoke.

Peter had seen, and the sight had well-nigh overcome him. Both boys stopped, and with ghastly faces gazed at the grewsome sight. On the ground near the fire were strewn human bones, and the meaning of the feast was only too apparent.

Suddenly Father Roubaud appeared among them. He was a large man, and his broad face had usually borne an expression of calmness that had been marvellously comforting to the boys, and apparently

not without a restraining effect upon the impulsive
red men to whom he ministered. But now the good
man seemed to be beside himself with indignation.
His face was purple, and the great veins stood out
upon his forehead.   Striding into the very midst
of the assembly he faced the men boldly, and began
to speak in tones expressive of the utmost wrath
and the sternest rebuke.

The words of the priest were received in silence,
and even when he ceased, no response of any
kind was made.   The men, both those who were
seated and those who were standing, remained like
so many carved images.   Not even at one another
did they glance, and so far as any visible effect
was to be seen, Father Roubaud might as well have
spoken to the giant trees of the forest.

After waiting in vain for a response, the priest
began again, this time in changed tones, when
pleading had evidently taken the place of the
sterner words which he had used before.   His voice
was tender and low, and it soon had its effect upon
his dusky hearers.

When he ceased one of the savages arose and
in his mildest tones began to speak to their friend
the priest.   His manner was calm and deliberate,
and the expression on his face was by no means
harsh or hard ; but it was manifest that they were
not disposed to heed the protests, and were determined to continue the sickening feast.

Father Roubaud again pleaded with his "children"; but though his words were received in a not unfriendly silence, they were unavailing, and at last, with a sob breaking from his lips and his hands at his eyes as if he would shut out the horrible sight, the good man withdrew.

"Come! We'll go, too," whispered Peter.

Silently the boys turned, and by a circuitous way succeeded in gaining their wigwam once more. Hastily entering, they dropped the blanket curtain behind them and seated themselves upon the ground. Both were deeply shaken by the experiences through which they had just passed. To be shown that they were prisoners and were to be kept within the confines of the village, although the knowledge was not new, was still sharply to define their position, but even that was almost forgotten in the horror of the sight they had witnessed.

John was the first to break the depressing silence, as, looking up at his friend, he said, " I never knew that before."

" What ? "

" That the redskins were cannibals."

" I've heard of it, but I never saw it before, and I hope I never shall again. I remember asking Sam about it one time, and he said that some of the tribes were, but not all of them."

" Perhaps we'll be the next victims, Peter."

Peter shuddered as he said, "I hope not." Still, the thought had been in his mind, and the horror of the great fear could not be shaken off. "We'll have to find some way out, John."

"I don't know what it 'll be," replied John, gloomily.

"Neither do I. All I say is, that we must find some way, or make one. I'd rather die from a club or a bullet if I have to die, than — " Peter did not complete the sentence, but he knew that his friend understood.

In the midst of their conversation food was brought them, and not even their recent experiences could deprive them entirely of hunger, especially as the food was well cooked and there was an abundance of it.

Their spirits somewhat revived by their repast, they began to look upon their predicament in a slightly more hopeful light; but though they talked long and earnestly, no way of escape suggested itself.

In the afternoon they again ventured forth into the village, where it was once more made plain to them that though they were given the freedom of the place, they were not to be permitted to leave it, for again when they drew near the border they found their way barred by the presence of three warriors who quietly indicated by their gestures that the boys were to turn back.

For the two days following their experiences
were unchanged. Their wants were abundantly
provided for, food was given freely, they might go
when and where they chose within the village ;
but whenever they approached the border their
further passage was invariably blocked by the
sudden appearance of men near them. In the
night-time guards were stationed, and were much
more vigilant than in the day, as the boys learned
when one night, after midnight, they stealthily left
their sleeping-place and strove to enter the forest.
The warning was more sharply given this time,
and they were accompanied by the red men to the
very entrance of the wigwam.

The fourth day of their captivity had arrived,
and conditions still remained the same. They were
still well fed, granted a large degree of freedom,
ignored by the tormenting Indian boys, and left
almost to themselves, except when they turned
their footsteps toward the forest; then the never
failing barrier presented itself, and they were compelled to turn back into the village. It was all
strange, seemingly unaccountable. Why they were
held or for what purpose, the boys would not let
themselves conjecture, though in the heart of each
there was a growing fear that he might soon learn.
But neither spoke of his own suspicion to the other
until the night of the fourth day had arrived.

It was late, and the village was silent save for

the occasional barking of a dog, but neither of the captives had been able to sleep.

" Awake, John ? " whispered Peter, unable to endure the silence longer, and sitting erect as he spoke.

" Yes," responded John, rising also.

" We can't stand this much longer."

" That's so.　But what can we do ? "

" That's just it.　What can we do ?　We must do something."

" What ?　You've something in your mind, I know.　What is it, Peter ? "

" We can't get into the woods."

" So it seems."

" I wonder if we couldn't work it on the shore."

" How ? "

" If we could only get a canoe hidden somewhere in the daytime we might be able to get to it at night.　They don't seem to be on the lookout for us so much on the shore as they do by the woods."

" We haven't tried it, have we ? "

" Not much.　Perhaps that's the very best reason why we ought to try it now.　We can go down there to-morrow, anyway, and see what we can see."

" I'm agreed.　I'll do anything you say, Peter."

" We'll try it, then, to-morrow morning.　If we can only get a canoe where we can find it at night, I believe we can get away.　It's worth trying, anyhow."

"So it is. I'm with you, Peter," said John, hope beginning to rise in his heart.

"Go to sleep now or you won't be in shape for anything —"

Suddenly Peter stopped as the blanket in front of the wigwam was lifted from its place and the figure of a man could be seen standing in the dim light. Instantly the boys cast themselves upon the ground, overwhelmed by the fear that the sound of their whispered words had been heard, and that the ever present warriors were at hand to repress them.

The man in the entrance had not moved from his position when Peter looked again, and this time he recognized him as the priest who had brought them to the village.

"It's Father Roubaud," he whispered to John, sitting up again as he spoke.

"His-st!" came the low warning from the entrance, and the visitor stepped inside the wigwam, dropping the blanket into its place as he did so. The room was now in total darkness, and the stillness was as intense as the darkness, while the trembling captives waited for the explanation of the unexpected visit.

CHAPTER XXIX

A Pathless Forest

PETER could feel rather than see that Father Roubaud was approaching the place where he and his companion were seated, and instinctively the two boys arose. The manner of the priest's entrance, as well as his stealthy approach, convinced them that something of more than ordinary interest was occurring, and they breathlessly awaited his coming.

Then Peter felt a hand laid upon his own, and though no word was spoken, the gentle pressure indicated that he was to follow. He stopped only to whisper to John, "Come on," and the three began to move silently toward the exit. There was no hesitation on the part of their leader when the blanket was lifted from its place and they stepped forth from the hut. Father Roubaud glanced keenly about him, and then began to move swiftly toward the border of the village, the boys following obediently, and wondering what it all meant.

When they arrived at the edge of the woods not a watcher was to be seen. The stars were twinkling in the clear sky, and though there was no

moon, the sleeping village and the giant trees stood
out plainly in the dim light. The priest halted only
long enough to convince himself that they had not
thus far been seen, then entered the woods, the
boys still keeping close behind him. They had advanced carefully for what appeared to Peter to be
a quarter of a mile, when the good man stopped.
Into Peter's hand he thrust a small object which
proved to be a compass, then without a word he
left his companions and turned back toward the
village.

It was all plain now, and Peter turned to his
friend, saying in a low voice : —

"We're free to go now, John."

"Looks like it," whispered John in reply.

"He's a good man."

"That's what he is. Can we ever do it, Peter?"

"Do what?"

"Find our way back to Fort William Henry."

"We've got to find it."

John was silent as they started. Neither voiced
the thought that the first thing to do was to place
the greatest possible distance between them and the
village from which they had just departed. Pursuit was the great source of fear, and prompt action
on their own part would render that less likely to
succeed. Accordingly, at a swift pace they kept
on their way till the stars became dim and the
eastern horizon was all aglow with the light of

the coming day. Then, for the first time since they had parted with Father Roubaud, they halted and looked at each other.

"It's great, Peter!" declared John, enthusiastically. "They'll never get us now."

"Not 'now,' perhaps; but we aren't in the fort yet, by any manner of means. What are we going to have for breakfast, John?" went on Peter, soberly.

"We'll find something. And even if we don't, it 'll be better than to have those redskins make a breakfast on us. Wasn't that horrible, Peter?"

"It certainly was."

"Have you any flint and tinder?"

"Not a bit."

"We might make a fire with sticks. I've heard that the Indians do that by rubbing two together till one takes fire."

"I've heard it too, but that isn't the same as being able to do it."

"We can try it, anyway," said John, at once beginning to look about him for some sticks to use.

"Hold on, John!" called Peter. "I'm not sure we ought to have a fire, even if we can make one. We probably aren't more'n twelve or fifteen miles from the village yet, and if they're on our tracks, it 'll be just like sending 'em word where we are. We'd better wait before we try it."

"You're right, Peter. But come on! Let's put

a few more miles between us. We're good for that, even if we don't have any breakfast."

"We'll lie down here and get a little rest first. It 'll save time in the end."

Both boys threw themselves upon the ground and remained without speaking until more than an hour had passed. Then at Peter's word the flight was resumed, and they continued on their way until the sun was high in the heavens. Once more they halted, weary now, and somewhat faint from the lack of food.

"We've just got to find something to eat," said John.

"What 'll we look for?"

"I'll tell you. Here's this brook," and as he spoke, John pointed at the small stream on the bank of which they had halted. "It's fairly alive with little fish. You can scoop some up while I'm rubbing the sticks together and starting a fire. Oh, I can do it! Don't you be afraid!" he added lightly, as he perceived that his friend was somewhat incredulous. "You 'tend to your part, and I'll do mine. And it won't do any harm to have a fire either, for we're a good long way from the village."

"But there may be other villages, you know, John. I've been on the lookout all the way for fear that we should run plump into one."

"That's so. Strange I hadn't thought o' that.

But then, Peter, we're too far in from the shore for that," he hastily added.

"Perhaps we are," assented Peter, though he still hesitated. "All right. I'll try it; but keep your eyes and ears open, John. We haven't a gun or anything to use if we're attacked."

"Shan't need it in that case," remarked John, simply.

While John was selecting material for his fire, Peter went farther up the little stream, and finally coming upon a shallow basin into which the waters fell, he decided that he would attempt to secure their breakfast there. The water certainly was swarming with little fish that darted away at his approach but speedily returned, having as yet but slight fear, so unaccustomed were they to the sight of a man. The stream itself tumbled merrily forward on its way, and the warm air of early summer, the almost cloudless sky above, the great trees that could be seen far as the vision extended on either side, the chattering squirrels, the songs of the birds that came apparently from every bush, the soft rustling of the leaves and branches, all combined to impress Peter with the beauty and majesty of the primeval forest. Peace lay within its borders, and here, to all appearances, there was nothing to be feared.

Yet Peter Van de Bogert clearly understood how deceptive all these things were. He and his friend were trying to escape a deadly enemy, the more to be feared because he was unseen. At any moment, from the midst of the peaceful scene upon which he was gazing, might come the report of a rifle, or the wild yell announcing the approach of the painted savages, terrifying in their aspect as they were relentless and vindictive in war.

Nervously Peter looked about him, but saw nothing to increase his alarm, and he bent low and dragged the trunk of a fallen tree to the border of the brook, letting it fall directly across the stream into the water. There was a scurrying of the little fish, and the clear waters were muddied; but unmindful of the disturbance he had aroused, Peter brought some of the larger stones from the bank and cast them into the water beside the log until he had made a fairly good barrier. Then he stepped back and waited for the water to become clear again and for the frightened fish to return.

As soon as this was accomplished, cap in hand, he stepped into the little brook, several yards above his dam, and began to move down the stream, driving the myriads of frightened fish before him. Reaching the obstruction, he began to scoop up the fish in his cap and to fling them out upon the grassy bank. The operation was repeated several times, and then, assured that he had secured all that would be required for the meal, he cast the smaller ones back into the brook, and cut "stringers" from the

near-by willows upon which he "strung" the larger
ones of his catch.   Then, his part of the task completed, he set forth on his return to the place where
he had left John.

As he came near, he was not able to detect any
rising smoke, and when he saw John, he perceived
that as yet he had not been able to secure any
fire.

"Any luck, John?" he inquired.

"Not a bit!   Can't get the stuff even to smoke!"
John's face was dripping with perspiration, and its
expression betrayed rising anger.

"Let me try it," said Peter, as he cast his string
of fish upon the grass.

But Peter's efforts were only slightly better rewarded than his friend's.   He did succeed in obtaining a tiny curl of smoke, but the branches would
not ignite.   By turns the boys continued their
efforts, but failure was still theirs.   At last, utterly
wearied, John exclaimed : —

"We can't do it, Peter, and we might as well
own up to it."

"No," admitted Peter, reluctantly, "we can't do
it."

"What shall we do?"

"There's the fish."

"Eat 'em raw?   Is that what you mean?"

"Yes."

"Faugh!   I don't want any."

"It may be better than nothing," suggested Peter, seriously.

"Not for me!" replied John, nevertheless taking up one of the fishes and lifting it to his face. But with an exclamation of disgust he threw it far out among the trees.

"Let's start on. We've been here too long already. Perhaps we'll find something on our way," said Peter, rising as he spoke.

Without a word of protest John followed him, and in silence they held to their course until the middle of the afternoon. They were utterly wearied and wretched now, but though both had kept a careful lookout for food, neither had discovered anything. It was too early for berries, and they were not as yet ready to resort to the bark of the more tender tips of the trees. Both boys slept for an hour and were somewhat refreshed when they awoke and prepared to resume their journey.

A bird's nest was discovered soon after they started, and the five eggs which it contained provided a little nourishment, but when night drew near and they began to look about for a sleeping-place (for they had decided to rest until morning), not another mouthful of food had been found. As if to make matters still worse, there was a heavy shower in the night, and the boys were thoroughly drenched. By daylight they were again on their

way, though their progress was much less rapid
now, and it was seldom that either spoke. The
storm had passed, but the brightness of the morning found no response in the hearts of either of
the fugitives. Though they frequently broke the
tender branches from the trees or bushes, and
chewed them, there was slight nourishment to be
obtained in this way, and their weakness increased.
John occasionally staggered, and the necessary stops
for rest became increasingly frequent. There was
an expression, too, in his comrade's face that was
causing Peter much anxiety, and made him to a
degree unmindful of his own sufferings.

It was on the fifth day of their journeying that
Peter had his hardest experience. On the preceding day he had compelled himself to eat of
some raw fish which he caught in a little brook
in a manner similar to that which he had employed
on the former occasion. John, however, had eaten
but little, even his fierce hunger not being sufficiently strong to compel him to partake of his
friend's repast. An occasional bird's nest had been
found, and the eggs, together with bark of the
most tender tree twigs and a few succulent grasses
that Peter had found and recognized, had provided
nearly all the food which the young provincials had
secured.

Their pace, too, as was but natural, had become
much slower, and their halts much more frequent.

The sole source of comfort had been that not once had they seen the form of an enemy. To every appearance the sombre forest was deserted of all save themselves. At night, when they had sought some secluded spot for rest and shelter, they had not heard even the sound of prowling wild beasts.

With the passing days the fear of starvation grew sharper, and as Peter perceived the steadily increasing weakness of John and saw how thin and pale his face was becoming, his heart sank within him, and a nameless fear arose to add to his own sufferings. Still he had urged his companion forward, striving to appear strong when his own weakness almost overpowered him, and it had seemed as if he too must give way under the strain, as John was certainly doing.

It was on this fifth day, however, as has been said, that the climax came. They had resumed their march in the early morning and had been an hour on the way, when John suddenly stopped, and looking earnestly before him exclaimed, "Do you see her, Peter?"

"Who?"

"Why, Lucy Locket! There she is right behind those alder bushes. Lucy, I've found your pocket!" he shouted, and, laughing loudly, began to run toward the bushes to which he had been pointing.

"John! John! Come back!" called Peter, as

he quickly followed his friend, who now had disappeared from sight.

"John! John!" he called again, despairingly.

For a moment he stopped, as the words of a song could be faintly heard far in advance of him.

"Lucy Locket lost her pocket,
Kitty Fisher found it.
Ne'er a penny was there in it,
'Cept the binding round it."

CHAPTER XXX

Lucy Locket's Pocket

FOR a moment the startled boy stood still, his pale face becoming still paler and his body trembling with excitement and fear. The wild laugh in which the words of the song ended, the shrill and unnatural sound of John's voice, and the fact that his companion had of late disregarded all the caution which they had observed throughout their flight, filled his heart with a feeling of alarm. Something was wrong with John, but what it was he hardly dared to whisper even to himself.

He darted on into the forest, following as best he was able the direction from which the song had come, for John's voice could no longer be heard; but for several minutes his attempts to find his friend were unavailing. At last, however, he perceived him kneeling upon the ground and looking earnestly upward as if he were watching some object that terrified or fascinated him.

Mechanically, Peter, too, stopped and gazed in the direction in which John was looking, but he was not able to discover anything to increase his alarm or explain the strange actions of his friend. Approaching the place where John knelt, he said : —

"John, what is it ? "

"Don't you see ? " replied John, without turning away his eyes from the place which held his attention.   "Don't you see it ? "

"I don't see anything.   What's wrong ? "

"Look at it !   There it is right up there in the branches of that maple."

"I can't see anything except what ought to be there."

"Then look again.   It's right before your eyes. A blind man could see it."

"See what ? "

"Why, Lucy Locket's pocket."

John was speaking quietly and in his ordinary tones, but a shiver passed over Peter's body as he listened to the words.   He placed himself where he could see John's face, and his heart was heavy when he perceived the wild and unnatural expression in his friend's eyes, which seemed almost to be starting from his head.

"John," said Peter, placing his hand on his friend's shoulder, and endeavoring to speak soothingly, "there's nothing there."

"Yes, there is something there, too !   I can see it, and so can you if you'll look where I tell you to. It's right on that long branch, to the left."

" There isn't anything but the leaves. That's what you see."

" Leaves! Leaves!" retorted John, scornfully. " I tell you it's Lucy Locket's pocket. You heard she'd lost it, didn't you?"

" Yes."

" Well, there it is, right before us. Poor Lucy! I'd never dare to look her in the face again if I didn't get it for her when I've such a good chance as this."

" But some one found it for her."

" Who found it?" John turned his face toward his friend as he spoke, and Peter was hardly able to repress the sob that rose to his lips as he looked upon it. It had the expression of a man no longer himself, the vacant look of one whose mind had fled.

" Why, don't you remember? Kitty Fisher found it for her," he said soothingly.

" Kitty Fisher? Kitty Fisher? Who's she, I'd like to know?"

" Why, she's Lucy Locket's friend, and found her pocket for her."

" She didn't give it back to her, if she did find it."

" I think she did."

" How could she, when it's there on the limb of that tree?"

" We'll go up and see if it is really there," sug-

gested Peter, not knowing what to do, but deem-
ing it best to humor his friend's vagaries. "Come
on, and if we get it we'll take it right back to
her."

John at once acted upon the suggestion, and
rising, began to run toward the place, Peter keeping close by his side. He was determined not to
lose sight of John again. There was no question
now as to what had befallen his friend. The pri-
vation and suffering through which they had been
passing, especially the lack of food and the continual anxiety of their flight, had at last had their
effect, and John's mind had given way. Peter
could not doubt it as he watched the frenzied
actions and saw the wild light in his eyes or heard
the incoherent mutterings as he breathlessly rushed
forward.

When at last they stopped beneath the overhang-
ing branch and John stood staring stupidly up into
the tree, Peter said quietly, "You see, John, it isn't
here."

"It was here just a little while ago."

"Perhaps, but it's gone now."

"Where? Where?" demanded John, his excitement increasing.

"Why, don't you know?" replied Peter, a sudden inspiration seizing him.

"No, where is it?"

"Kitty Fisher took it."

"She did? She did? Where is she? I'll make her give it up. She has no right to it."

"So she hasn't. We'll make her give it up."

"But where is she? You haven't told me that."

"I don't just know, but we'll start right off and we may find her."

"Come on! Come on, then!" said John, abruptly darting into the forest.

"Not that way! Here! Come with me!"

Peter hurriedly turned in the direction which he hoped would lead them on toward Fort William Henry. He had purposely made a wide detour on their way, hoping to go far enough inland to avoid the danger of being taken by any of the men near Ticonderoga. His one hope now was to induce his companion to follow him, and if John could bear up for another day or two, he could get him into the place where proper care would be given.

John was comparatively docile, and at Peter's sharp call had turned and was following him obediently. He frequently muttered to himself, but apparently his desire to secure the lost pocket was gone, for as they journeyed on he made no further reference to Lucy's loss. He was weak, however, very weak, and Peter's eyes were at times filled with tears as he saw him totter or perceived how thin his face was and how emaciated his body had become. It was pathetic beyond expression. Frequently he stopped to enable his friend to rest and to rest himself, for Peter had to admit his own increasing weakness. If the end of the journey were not speedily achieved, then never again would he and his friend be within the shelter of the fort. To go on much longer was impossible, but to give up now was to give up all. He must keep on, he told himself with set lips, although every time the journey was resumed his own weakness became more apparent. Darkness came at last, and too exhausted to use the caution he had up to this time maintained in seeking a secluded place in which to sleep, Peter with his companion almost fell into a clump of bushes near their pathway and stretched themselves upon the ground.

A heavy rain fell in the night, and both boys were stiff and sore when they rose in the morning to continue their journey. To Peter it almost seemed as if he could go no farther. Strange lights were dancing before him and strange sounds roared in his ears. He was somehow aware that John was unusually quiet and docile and made no protest when the journey was resumed; but he himself was stumbling, and frequently fell to the ground, rising each time with increasing difficulty. Not a mouthful of food had either had for breakfast, but the morning sun was warm and the air soft and balmy, almost their sole comfort as they struggled onward, but one of which even Peter was only partly conscious.

All thoughts of time were lost, but there was
still in his mind a vague realization of the destination they were seeking, and he frequently glanced at
the little compass which Father Roubaud had given
him on that eventful night when he had led them
forth from the Indian village.   But that experience
belonged to a remote past into which the young
provincial's thoughts were hardly able to enter.   It
seemed to him that they had for years been doing
nothing except stumble on through an unbroken
forest, searching for they scarcely knew what, and
ever on their guard against an enemy that never
abandoned the pursuit.

It was full noon and the heat had increased.
Both boys were so utterly weakened that they were
hardly able to move now, and Peter was about to
tell his comrade to stop again for the frequent rest
they were compelled to take, when he perceived
that John had already stopped and was smilingly
gazing before him, as if he had discovered something that greatly pleased him.

"Look, there she is!" he said, pointing as he
spoke to a place off to their right.

"Who?   What?"

"Why, there's Lucy.   Lucy!" he called in a
faint, shrill voice, "I know where your pocket is!
But there's nothing in it, not a penny.   Just the
binding round it."   John was laughing and child-
ishly clapping his hands; but Peter hardly knew

what his friend was doing, for in the place to which John had pointed he beheld the forms of several men, yes, and there were women, too, in the number. Whether they were friends or foes he could not determine, and somehow he seemed to be strangely indifferent. He did not call, or even attempt to run, but, overcome by his excitement, sank to the ground.

He was conscious that the people were approaching, he could still hear John's occasional reference to Lucy Locket; but who the strangers were or what their purpose was he did not know or even care to know.

What followed he never was able to recall clearly. He felt dimly that he was being assisted to move and that the party was advancing through the forest, but whether the French fort or the English was to be the destination he did not understand. He could do no more, and the end of the long journey had been gained; though what that end was, to him seemed a matter of supreme indifference.

Not even when he was conducted at last into a building of some sort did he comprehend what it was. He heard the sound of voices, he knew that he was being placed in a bed, and that he was bidden to drink the contents of a wooden bowl which he was too weak to hold to his own lips. And then he fell asleep, without a thought of what had become of John or whether they were in the hands of enemies or friends.

When at last he awoke it was again broad day-light, and a woman who was in the room with him once more held the bowl to his lips and then bade him lie down and go to sleep. And the exhausted young soldier was too eager to comply even to stop to inquire where he was.

The next time his eyes opened he beheld Sam, the hunter, seated near his bed. Peter smiled, for the sight was a welcome one; and when more food was given him and he was again bidden to be quiet, there was the first glimmer of interest in his surroundings that he had felt for a long time.

"Sam," he said, "I'm in the fort, am I not?"

"Looks like it."

"Where's John?"

"Not very far away. Where you be is most likely where he is too."

"Is he all right?"

"I s'pect he is. He's in good hands, anyway."

"Who is taking care of him?"

"Sarah was. Now go to sleep again."

"Sarah?" Peter tried to sit up as he spoke, but fell back helpless.

"That's what I said. But I won't say 'nother word if ye don't lie still an' stop yer talkin'."

"But, Sam——"

"Here I go!" And resolutely the hunter departed from the room.

For a time Peter endeavored to conjecture what was going on in the fort, but only an occasional sound came to his ears. Nevertheless it was a source of deep satisfaction to know that he really had gained the refuge he had sought, and that John, also, was safe. He could not understand how it was that Sarah was caring for his friend, for she had distinctly told him when he had last departed from the fort that she was to return to her home. The problem was too difficult to solve, and at last he abandoned it. Then the recollection of the hunter and his presence was proof positive that he, and probably his companions, had also succeeded in returning safely to Fort William Henry. A feeling of contentment crept over him, and soon he was sleeping again.

Many days elapsed before the young soldier was able to leave his bed. The hunter was there at intervals, but he sturdily refused to talk, so that Peter was left in ignorance of what was going on. He did not know that the scattered families in the region had fled to the fort for shelter, nor even that it was one of these families that had brought him and John with them in their flight. He was likewise not informed that the army of Frenchmen and Indians had advanced, and that already Montcalm had sent a demand for the surrender of Fort

William Henry, a demand which Colonel Monro
had refused, confidently believing that General
Webb would send him aid from Fort Edward.
Sam and other scouts had been sent to that officer
again and again with messages from the colonel
begging for aid, and promises had been given, but
still no troops came. Sarah had been unable to
return to her home, as it was deemed unsafe to
make the attempt; and, besides, no one could
be spared from the fort to accompany her, for
peril was becoming greater daily. She had received no word as to the fate of her aunt and the
helpless children, or of Timothy Buffum.

CHAPTER XXXI

The Fall of Fort William Henry

THE August days had come before Peter Van
de Bogert and John Rogerson were able to
take their places in the ranks of the defenders of
the fort. But it was not the sight of the summer
sun nor even the pleasure over their own recovery
that took up their thoughts, for the hosts of Montcalm, numbering ten thousand or more, were advancing upon the place. The canoes of the Indians
in lines stretching even from shore to shore, the
heavier craft which bore the great guns and mor-
tars, the yells of the savages, the sound of muskets
as parties from the fort met those from the advancing army in the near-by region, were all-
absorbing. The charred trees and burned stumps
in the forest, the blue waters of the lake, the
hideous faces of the red men, the resolution expressed on the countenances of the defenders, and
the terror of the women and children under their
protection combined to make a picture which could
never be forgotten. Yet the growing desperation
of the situation only increased the determination
of the dauntless Colonel Monro and his equally

resolute followers to defend Fort William Henry to the very last.

Fourteen miles distant was Fort Edward, where General Webb was in command of twenty-six hundred men, for the most part inexperienced in war, though they were hardy and brave, as were most of the provincials. The peril of the neighboring fort was fully known by him, and he had been sending urgent messages to the governor of New York and to the New England people, explaining the threatening danger and begging that men be sent to his aid. Between Fort Edward and Albany were several posts; but to withdraw the defenders and send them with his own men to the help of Fort William Henry would be to leave all the region exposed to Montcalm's army, if the Frenchman should succeed in taking the places which first barred his advance. It is commonly believed, too, that General Webb was himself timid, hesitating, uncertain, and no fit man for the position he held; for his first and supreme effort should have been to strengthen the hands of Colonel Monro and prevent the enemy from seizing the fort at the head of Lake George. Some men came in response to his call, but they were few, and those he sent forward to Fort William Henry did not number more than a thousand all told. Their arrival increased the force to nearly three thousand, but surrounding them was an army of Frenchmen and Indians num-

bering more than ten thousand; and there is slight
cause for surprise that the heart of Colonel Monro
was heavy, although as yet he had no other thought
than that of defending the place, for he still believed
that aid would come from Fort Edward.

The slight skirmishing around Fort William
Henry soon gave place to much more serious work.
On the road which led to Fort Edward, Montcalm
had stationed a large force to intercept any re-
enforcements which might come from that place.
The French general, young, alert, and brilliant, was
directing the plans and movements of his men in
person. His first plan, of trying to carry the fort
by storm, was abandoned when he discovered that
the works of heavy logs and gravel were apparently
able to withstand his attempts; but he arranged
his forces with great skill, and when, as he confidently believed, he had the fort within his power,
he sent the following letter to Colonel Monro: —

"I owe it to humanity to summon you to surrender. At present I can restrain the savages and make
them observe the terms of a capitulation, as I might
not have power to do under other circumstances;
and an obstinate defence on your part could only
retard the capture of the place a few days and
endanger an unfortunate garrison, which cannot be
relieved, in consequence of the dispositions I have
made. I demand a decisive answer within an
hour."

In response to this demand, sturdy Colonel Monro replied, as might have been expected, that he would not give up the fort and that he and his men were prepared to defend it to the last. However, the doughty colonel fully realized his peril; and the sight of the multitudes of savages that, as soon as his response was known, came within open view of the fort and shouted their wild cries of defiance as they danced about, caused him to send another messenger to Fort Edward with a still more earnest and pitiful plea for help. This messenger, on his return, bearing a letter from General Webb in which Colonel Monro was informed that he must not expect aid from Fort Edward, was unfortunately taken by the Indians; and the letter which was found upon his person was at once turned over to Montcalm himself, who quietly held it for a use for which he believed it would shortly be required.

The French army dug trenches by night, and from these began to fire more frequently upon the fort. But the reply from the guns was vigorous and determined, and their continuous roar, as well as the sharper reports of the rifles and muskets, broke the stillness of the August days. Steadily the French army improved its position and drew nearer the walls. The Indians, like troublesome children, were often in the way and always in evidence. They begged to be permitted to fire the cannon, and laughed and clapped their hands with glee when the thunderous reply followed their touch.   They rushed from place to place, mostly without discipline or order, but the sight of them was far more frightful to the inmates of the fort than that of the disciplined soldiers of the French army.

At last Montcalm arranged what he thought was to prove his supreme action.   After a brief period of silence he meant to have his guns on all sides of the hardly beset little fort open fire together upon its walls.   The firing was to continue for a time, and then silence again was to ensue.

The plan was carefully executed, and in the silence which followed the deafening volley a messenger advanced into the open space before the walls, bearing a courteous, and doubtless sincere, word of praise from Montcalm for the determined and brave defence of the fort.   In the note was enclosed the letter which had been taken from the unfortunate messenger of General Webb, and in which the provincial commander had explained his own weakness, that the road from Fort Edward was held by Frenchmen and Indians, and that Colonel Monro must expect no aid from him.

What the brave colonel's thoughts were as he read the captured letter, enclosed without a word of comment in the complimentary note from Montcalm, can only be conjectured.   In response he simply expressed his pleasure in having to do with so courteous an enemy and so generous a man. And that was all. The siege and defence were still to go on. The energy of the attacking army now redoubled, and steadily it pushed its way nearer to the fort. The noise of battle increased. The whoops of the savages were loud throughout the day, and did not cease when darkness fell upon the land. Sorties were made from the fort, but the brave men were either driven back behind its walls or fell before the bullets of an unseen enemy. Volleys were fired, the roar of the guns continued, and stealthily and persistently, like the tide creeping up the sands of the seashore, the attacking army drew nearer and nearer to its goal.

Within Fort William Henry the condition of affairs was almost indescribable. Smallpox was aiding the Frenchmen, for many of the provincials were suffering from the dreaded disease. More than three hundred of the defenders had fallen in death or were lying helpless from their wounds. To make matters still worse, almost all of Monro's cannon were useless, some having burst, while others had been struck by the shot of the enemy. In places, too, the walls of the fort were giving way, and an easy entrance awaited the resolute and energetic young French commander, who, the colonel knew only too well, would be prompt to follow up his advantage.

It was now the eighth day of August, 1757.

Throughout the day and far into the night Montcalm's guns were speaking with a fury that, to the almost distracted commander of the fort, seemed to increase with every passing moment. The sickness among the people in the fort, the presence of the terrified women and children, the fact that the store of powder was becoming low, the knowledge that the walls were yielding before the terrible and unbroken onslaught, and that at almost any moment might occur an assault before which he and his men would be powerless, all combined to make the sturdy colonel almost desperate. All through the hours of that fearful night the few remaining small guns in the fort were kept steadily at work, but the end had almost come.

In the early light of the morning of the 9th of August, Colonel Monro summoned a few of his most trusted officers and set before them the exact conditions that faced them. The courage of the men was unbroken, but courage is no substitute for powder, and determination cannot take the place of cannon. It was decided at last that, if Montcalm would grant terms of surrender which could be honorably accepted, Fort William Henry should be given up to the Frenchmen.

Hardly had the decision been made before a white flag was flying from the battered walls of the little fort, the noisy beat of the drum soon silenced all other noises, and on horseback, accompanied by several of his men, Lieutenant-colonel Young sought the quarters of the victorious French commander.

His approach was greeted with silence, although the eyes of the curious savages as they watched the little party were eloquent in their way. The provincials were received within Montcalm's tent, and his reply was speedily made known and borne back to Colonel Monro, waiting behind the walls of Fort William Henry.

In the flush of success, a success due primarily to the carefulness with which Montcalm had prepared and carried out his own plans, and to the lack of this very care on the part of those who were most responsible in the English provinces, the young French general showed himself a great man. Respecting thoroughly the bravery of Colonel Monro and appreciating also the valor of his troops, he agreed that the provincials should receive all the honors of war, and should also be permitted to go to Fort Edward, being escorted and protected on their way thither by a detachment of French troops. This last condition had seemed to Montcalm unnecessary, but he was soon to learn that Colonel Monro had not asked for the attendance of the French troops without having good ground for the request.

On the other hand, the provincials were to promise that they would not enter the service of the

English provinces again within a year and a half at the least, and they were also to restore all the French prisoners that had been taken on American soil since the beginning of the war. Naturally, and justly, Montcalm insisted upon all the stores and guns of the fort becoming his, but he generously declared that his foes should keep one of the field-pieces as a reminder of the brave defence they had made before the fort was surrendered.

Montcalm's first act was to summon the leading chiefs of his Indian allies, explain to them the terms of surrender, and urge upon them the necessity of restraining their warriors when the crucial moment should arrive for Colonel Monro and the occupants of the fort to depart. With one accord the red men agreed to do what Montcalm requested, but he was soon to learn that a promise and its fulfilment were by no means synonymous in the thoughts of his savage allies.

A guard of French troops had been stationed near a camp which had been made and intrenched not far away, and in this camp the entire body of men, women, and children took refuge when they left the fort. They had been compelled to leave behind them those who were too badly wounded or too ill to accompany them, and the first sign of trouble came when a rabble of yelling Indians plunged into the abandoned fort and instantly put to death all the wretched occupants. Eager for plunder, their next action was to seek for such articles as had not been taken with the fugitives. They found little of value, and the failure increased their rage. They rushed from the fort and began to seek among the tents for what they desired; they seized the simple ornaments from the terrified women; they demanded rum from the flasks of the men, and with the thought of avoiding greater peril, the flasks were given to them. Pushing against one another, giving vent to their blood-curdling cries, brandishing their tomahawks, the painted warriors at last forced their way past the soldiers and into the heart of the camp into which all the fugitives had been assembled.

CHAPTER XXXII

CONCLUSION

THROUGHOUT the long night the savages brought terror to the frightened and defence-less people huddled together in the camp. The men, it was true, had been permitted to retain their guns, but they were without ammunition, and had no bayonets. Even if these things had been in their possession, it is doubtful if the provincials would have made use of them for fear of increasing the rage of the Indians, who were already beyond restraint. It was a night of terror, and when at last the morning dawned, the fear of what lay before was even more distracting than the memory of that which had passed. Whatever trinkets were seen on the persons of the women or children were snatched away by the brutal hands of the savages; and screams of terror were frequently heard as some painted warrior seized a woman by the hair, and, with unmistakable gestures, demanded her pos-sessions. The cries of little children were almost continuous, and not a captive had dared to close his eyes throughout the summer night.

When morning at last came, the report that

others of the sick or wounded soldiers had fallen victims to the fury of the Indians increased the fear of the assembly to such a degree that Colonel Monro demanded of Montcalm the observance of the terms of surrender and the protection which had been promised.

The young marquis was almost as desperate as his prisoners. Already he had sent men to Montreal with the news of the fall of Fort William Henry, but the feeling of elation over his victory had given place to consternation and sorrow as he perceived how powerless his men were to restrain their savage allies. He promised the colonel to do his utmost, and at once assembled the leaders of the red men, begged them to assign two chiefs from each tribe to accompany the escort which was to conduct the prisoners to Fort Edward, and urged them to restrain the red men under their charge from molesting the defenceless. This was readily agreed to, but even Montcalm must have been fearful of the result when at last the march toward Fort Edward was begun.

His worst fears were realized. Elated over their success, and already wild from the ease with which their own desires were fulfilled, the painted warriors soon began to crowd into the lines, and the scenes that followed are too horrible for description. As the march proceeded, matters steadily became worse. The officers of the French pleaded,

begged, and threatened in vain. Occasionally, the
sturdy Englishmen made a stand and endeavored to
defend themselves and the helpless women and children who were with them, but their actions only
served to increase the wild excitement of the warriors and to make their own position worse. The
Frenchmen begged of them not to excite the Indians, but to hasten forward on their march and
gain the shelter of Fort Edward in the shortest
possible time.

Yet fourteen miles were to be traversed before
the desired shelter could be had, and with every
passing moment the frenzy of the savages increased.
To-day, as we look back upon the terrible scene of
the soldiers and women and children trudging along
the rough road that led through the forest to Fort
Edward, the wonder is that a general massacre did
not follow. As it was, even the French officers,
who afterward were inclined to belittle the reports
of the tragedy, admitted that they had counted as
many as fifty dead bodies left on the line of march,
and we may be certain that there were many more
who fell victims to the frenzy of the Indians on
that August day.

Montcalm was almost beside himself with sorrow
and anger. Rushing into the midst of the howling
mob, he shouted, "Kill me, but spare the English
who are under my protection!" With his own
hands he seized from the grasp of the Indians help-

less men or women and set them free. But even
his efforts were without avail to stop the wild disorder. In the midst of the shouts of the frenzied
savages could be heard the cries of women calling
for their children, wives for their husbands, and
friend for friend. Many of the terrified people
broke from the mob and fled into the forest, and
for three days the straggling fugitives came, singly
or in detachments of two or three, into Fort Edward, to which place they were guided by the
sound of the cannon fired at regular intervals, to
show in which direction safety was to be sought.

The column at last was completely broken up.
Montcalm and a few of his officers managed to
collect some of the scattered people, though many
had already fled back to the fort from which
they had departed, hoping to find shelter and
protection there, while others had returned to the
camp. Many more had escaped into the forest,
preferring to trust themselves to the danger of the
wild beasts or endure the sufferings they must be
compelled to undergo there, rather than be at the
mercy of so treacherous a foe.

Several hundred were at last brought together into
the camp, and Montcalm did his utmost to provide
for their wants and to insure their safety. With their
own money some of the French officers redeemed
articles of clothing which the Indians had seized,
and restored them to their owners. Food also

was provided and a strong guard set for defence,
and there the fugitives were kept until the fifteenth
day of August arrived.

Most of the Indian warriors departed on the day
that followed the attempt to conduct the prisoners
to Fort Edward, and with their departure a
measure of confidence and safety was restored.
On the 15th, under the protection of an escort
much larger and stronger than that which had been
given previously, the remaining prisoners were conducted to Fort Edward, where they arrived safely.

Of the events that followed in that place, the
limits of this story forbid a full description.
There were scenes never afterward to be spoken
of calmly by those who witnessed them. Families
had been separated in the wild flight, and when
mothers found their children, or wives their husbands, among the fugitives that came straggling
into the fort, the pathetic reunion beggars description. There were others who waited in vain for
the coming of their loved ones. It was said that
the Indians in their flight to Montreal had taken
with them more than two hundred prisoners.
Some of these returned later to their friends,
but of others no word was ever received.

Of those with whose fortunes this story has
been more immediately concerned only a brief
word can be given. Soon after the sound of the
wild war-whoop which indicated that the red men

were about to fall upon the straggling column moving toward Fort Edward, Peter Van de Bogert found himself near his friend Sarah, just as she escaped from the grasp of a huge warrior, who had seized her by the hair with one hand while with the other he tore away a simple brooch from her neck.

At Peter's word she had followed him into the forest, nor did they cease running until the fearful sounds could no longer be heard and they felt comparatively safe from pursuit, for there were too many left behind upon whom the Indians could bestow their attention to permit of following far those who fled from the lines.

Peter, who was thoroughly familiar with the region, had but little difficulty in making his way to Fort Edward, and he and his friend were among the first to gain its shelter. There Peter remained until the summer was ended and he was once more permitted to return to his home, whither Sarah, in company with numerous others, had been sent in the course of a few days.

John Rogerson, also, had been among the number that gained the shelter of Fort Edward; and there he too remained until the summer was gone, when he returned to his New Hampshire home. But neither of the young provincials had as yet seen the end of the struggle between France and England for the possession of the New World. Roused

at last by repeated reverses, Old England began
to realize something of the task which was hers.
Men were raised and leaders were found competent
to cope with the difficulties, but of the success
which was finally won the present tale has nothing to do.

Of the fate of Fort William Henry itself, a word
is necessary. On the day following that in which
the wretched prisoners were safely conducted to
Fort Edward (August 16, 1757), Montcalm's army
was busied in destroying what remained of the
fort. The buildings and barracks were torn down,
and the timbers cast into one pile. To this pile
were added the logs from the ramparts, the bodies
of the dead were placed upon the huge pyre, and
soon a mighty blaze lighted up the quiet waters of
the lake. All night long the great fire burned, and
when the morning dawned only the smoking embers
remained to mark the site. The fort was never
rebuilt. To-day, to one who stands on the spot
and surveys the beautiful scene before him, the
clear waters, the towering hills, the green-clad
meadows, the wooded slopes, and the peaceful
landscape make it seem rather a dream than a
fact that here was enacted one of the most terrible
tragedies of the New World. And yet the dream is
true. The waters were covered with countless
canoes, the hills reëchoed with the thunder of
great guns, and here men met in a deadly conflict

for the possession of the land. It is impossible to estimate its present worth or appreciate its present peace without some knowledge of the days that are gone.

* * * * * * *

The scene of this tale shifts once more to the little hamlet in which Peter Van de Bogert was staying in the house of his aunt and his grandmother. Snow covered the land, and the air was crisp and clear. Peter himself had responded to a rap at the low door, ready to welcome the unknown visitor, whoever he might be, for visitors were so rare that their coming was ever a matter of interest. As he flung back the door he stood face to face with a man who gazed at him smilingly, and without waiting for a word of welcome stepped inside the room.

"Sam, it's you!" exclaimed Peter, eagerly, as he closed the door and stood before his friend.

"Seems so," laughed the hunter, "though that's as may be. I've been through so much I'm not jest sure o' it myself."

It was not long before Sam was seated at the table upon which dinner was already waiting, and the two fell to talking busily. There was a smile of content on Sam's face, and it was evident that he was as much pleased at the meeting as was Peter himself.

"Sam," said Peter, at last, "it seems like an age

since I've seen you. You never told me how you got back to the fort that time when we went down Champlain to set fire to the stores of the Frenchmen."

"Didn't I? Well, it's not my fault. Ye've never been 'round where I was. We all came back in our canoes, that is, except when we carried 'em overland."

"You got back all right?"

"So it seems. I b'lieve we was chased a few times, an' did have a brush or so with the redskins, but we fooled the Frenchmen good."

"Not all the time," said Peter, thoughtfully.

"*We* did. We didn't have any General Webb 'long with us, ye see. Why don't ye ask me where I've come from now?"

"Where did you come from, Sam?" inquired Peter, obediently.

"Montreal."

"Have you been there?"

"That's what I have."

"What for?"

"T' get Sarah's aunt an' th' children."

"Did you get them?" demanded Peter, jumping up in his excitement.

"I did. That's what I went for, wasn't it?" responded the hunter, quietly.

His story was soon told. The woman and her children had not been ill treated and had been carried to Montreal, where the Indians received a reward for every prisoner taken. The hunter's search had been long and difficult, but at last had been rewarded with success, and their release had been obtained.

"And was Timothy Buffum there too? Did you get him?" inquired Peter.

The hunter shook his head. "Couldn't find hide nor hair o' him."

"What became of him?"

The hunter made a significant gesture, but did not explain, and Peter understood.

The conversation was interrupted by another rap on the door, and Enos the postman was admitted. He slowly unrolled the huge muffler, his own handiwork, from his neck, and as he reached into his pocket for the letter he was to deliver, he suddenly noticed the hunter's presence.

"I d'clare!" he exclaimed, as he held forth his hand, "I'm glad t' see ye, Sam, but I didn't think I'd ever set eyes on ye agin. I didn't, honestly."

"I'm sorry t' disapp'int ye, Enos," laughed the hunter.

"No disapp'intment. Not a bit. When d'ye get back?"

"T'-day."

"Ye don't say! I s'pose ye been fighting Frenchmen all th' time."

"Not quite."

"He's brought back Sarah's aunt and the children," explained Peter.

"Ye do-o-n't say! I must go right straight home an' tell my wife. She'll be powerfully glad t' hear o' it."

"Don't go, Enos," said Sam.

"I've jest got to. I—"

"But I want to hear about that snake."

"Which one?" replied Enos, instantly abandoning his purpose to depart.

"Why, the one what—"

"The one that showed us where the gray squirrels hid their beechnuts?"

"Yes. Yes—that's the one."

"That's too bad, Sam," said Peter, in a low voice. "Come on with me."

"Where ye goin'?"

"I thought I'd go over to Sarah's. I want to see her aunt and the—"

"Oh, ye do, do ye? Well, ye don't want me 'long then."

"Yes, I do, Sam."

"Well, I'm not goin'. I've got t' stay here an' hear 'bout Enos's snakes. You go on, lad," he added cordially. "P'r'aps I'll be over later, but jest now I've got t' hear 'bout this story."

Whether it was the quizzical expression on the hunter's face, or his own eagerness to welcome the returned members of Sarah's household that influenced him most, Peter did not stop to consider, but hastily departed. As he came near the home he sought, he could hear the sounds of laughter within, and paused as he listened. It was good to have escaped all the perils of the past year, good to have the lost restored. The year itself had been filled with disasters, but perhaps the end of the struggle would be like the close of the eventful year. At least the hope was strong in Peter's heart as he advanced again, and, giving a loud rap, but without waiting for the formal word of wel come, opened the door and entered the house.

Printed in the United States
129107LV00006B/13/A